Nonlinear Smoothing and Multiresolution Analysis

Carl Rohwer

Birkhäuser Verlag
Basel · Boston · Berlin

Author:

Carl Rohwer
Department of Mathematics
University of Stellenbosch
7602 Stellenbosch
South Africa

2000 Mathematics Subject Classification: Primary 00A69, 41A46, 42C40; Secondary 06, 47, 62, 65, 94

A CIP catalogue record for this book is available from the Library of Congress, Washington D.C., USA

Bibliographic information published by Die Deutsche Bibliothek
Die Deutsche Bibliothek lists this publication in the Deutsche Nationalbibliografie; detailed bibliographic data is available in the Internet at http://dnb.ddb.de.

ISBN 3-7643-7229-X Birkhäuser Verlag, Basel – Boston – Berlin

Part of Springer Science+Business Media
Printed on acid-free paper produced of chlorine-free pulp. TCF ∞
Printed in Germany
ISBN-10: 3-7643-7229-X
ISBN-13: 978-3-7643-7229-3

9 8 7 6 5 4 3 2 1

www.birkhauser.ch

Contents

Foreword

This monograph is intended as a simple introduction to the so-called $LULU$-theory and the practical use of $LULU$-smoothers leading up to a full Multiresolution Analysis of any finite sequence. The attempt has been to present the subject in a way that is retrospectively ordered to some extent, but preserves some of the twisted paths that intuition initially suggested. It is based on the few publications in the subject, some already submitted and lectures given at conferences and on invitation at public and private institutions.

The ideas originated more than fifteen years ago when the author was involved in practical problems at the Institute for Maritime Technology in Simonstown and advanced while the author was at the BMI in Stellenbosch. The author hopes that publication of this book will stimulate the interest of other researches and takes responsibility for any remaining typographical or technical errors.

A growing love for this particular subject was one of the prime reasons for a return to academic life, and a perceived opportunity to consolidate various perspectives and results. It is however only the encouragement, guidance and help of several respected people that have resulted in the effort presented here. There is a deep gratitude for; Siegfried Göldner for teaching Mathematics, Gunther Meinardus for patience, good advice and courage, Johan de Villiers for researching with me in Spline Function, Vilmos Totic for pointing us towards Wavelets, Charles Chui for teaching these and Doron Lubinsky for trust, encouragement and walking many an academic mile. I also thank many colleagues and friends who have contributed by discussion or collaboration to the subject: Willie Conradie, Johannes Cronje, Tertius de Wet, Hans Eggers, Sven Ehrich, Frank Hampel, Trevor Hastie, Rolf Jeltsch, Henri Johnson, Gerhard Joubert, Odei Kao, Sergue Lapine, Dirk Laurie, Eberhard Malkowsky, Colin Mallows, Dana Murray, Lothar Reichel, Birgit Rohwer, Lourens Toerien, Johan van der Merwe, Marcel Wild, and the students that have attended my courses on the subject, especially J.P. Harper for his help with the sketches.

Finally, most of all, I thank Jan Malan for encouragement and support during the birth of the ideas, Lauretta Adams for her courage, patience and excellent word processing, Marguerite and our children for sharing me with $LULU$, and my parents to whom I dedicate this effort.

Introduction

> The order that I create, my father said, is the order of life. To create this order I shape a face permitting adoration. Order is a sign of life and not it's cause. Introducing life creates order, and introducing order co-opts death. Order for the sake of order is a parody of life.
>
> Antoine de Saint-Exupéry. *

Unlikely events do occur, but science, measurement and information are primarily concerned with tendencies, which we refer to in this book as "trends". Perhaps the most basic trend is monotonicity, as exhibited in the arrow of time. Tides and waves exhibit a form of local trend, if the minor local trends and large global trends are overlooked (smoothed out). Anomalies, characterized by an apparent lack of trend do occur, and are often inherently brief in space and time. When there is a lack of obvious trend, there is still possibly a statistical trend in the shape of a distribution.

Numbers and mathematics have evolved from measurement and economical information storage and transfer. If mathematics is considered the language of science, and the functions of a language are examined, it is reasonable to deduce an order in the functions (uses) as for example in the first four functions of a language as given by Popper (18). Corresponding to each function, an associated value comparison is apparent.

	Function	**Value**
1.	Expressive	Revealing/not revealing
2.	Signalling	Efficiency/inefficiency
3.	Descriptive	Truth/falsity
4.	Argumentative	Validity/invalidity

It is natural to consider higher functions (uses), like the poetic function, since even a strict formal language like mathematics can convey wisdom by using metaphors and analogies for modelling events, and leads to the power of interpolation and extrapolation as evident in the impressive accomplishments of modern science. Science develops from simple models of relatively constant things, including recurring

*Imprompto translated by the author from a German translation.

events of constant period or frequency, and builds up to more complicated trends. Increasing speed, accuracy and data transfer rates have often led to an overabundance of data from which a desired "signal" or trend is to be extracted in a stable and computationally economical way. An early example is the effort produced by Gauss to determine an elliptic orbit ("signal") from a sequence of measurements corrupted by undetermined and partly undeterminable perturbations produced by ray bending, instrument resolution, computational approximation, truncation and human error. A recent example is the transmission of information from an instrument on a distant planet. The relatively high data rate possible is an advantage but needs to be handled economically for good extraction of "signal", given perturbations essentially of similar type as those handled by Gauss. Two distinct changes are visible; The amount of work involved to extract the wanted signal and the type of noise regularly introduced by digital processing. Celestial orbits and the determination of their parameters, particularly with the limits of technology of the times, have been replaced by more complicated models for which parameters have to be estimated. Specifically, the advent of digital measuring, calculation and transmission of an ever increasing amount of data has led to problems that are not only different in detail but in philosophical basis.

Whereas extrapolation requires the use of well motivated non-local models, interpolation can use simpler local models for predicting values between measurements under reasonably mild assumptions on the behavior of a function. Spline functions have been a natural development linking global continuity requirements with local support in a computationally convenient way, permitting a well-known idea, the Raleigh–Ritz–Galerkin approximation for solutions of differential equations, to expand into a powerful tool called the Finite Element Method. Using computationally convenient interpolants by functions with local support, and comparing with the best global approximant with the Lebesgue Inequality, justifies the procedure in that it shows the error to be no greater than a small factor of the best possible in a given norm. Thus local procedures have been made respectable, and the art of numerical analysis has been saved from the suspicion of being a collection of clever ad-hoc techniques to calculate parameters in a complex model for the purpose of explaining or predicting events. What spline theory has done to link global and local analysis, wavelet theory has done to periodic events; localizing some frequency based phenomena coherently (8).

Intelligent ad-hoc local procedures for the removal of "impulsive" noise from data have received considerable attention in the last two decades. Wisdom and experience has resulted in the widespread use of the (running-) median smoother and compositions of these. A lack of clear and consistent behavior has been widely lamented, but experience indicates that these are effective procedures. Examples of problems from practice suffice to indicate the extent of the problem.

Suppose a sequence of numbers is transmitted with a reasonable trend in that all are measurements of light intensity at a pixel in a camera. If any one of the pixels now and then registers a "glint", or if one of the numbers now and then has a spurious influence in a significant bit, like a sign bit or exponent bit of the number, there is the annoying presence of "impulsive" noise. This noise is characterized by briefness and unreasonable size relative to neighboring data. Even though this type of noise was also present in previous times, human intervention and wisdom easily identified such an "error" and interpolated from neighboring data to leave a relatively undamaged signal. For high speed data processing the problem is to balance complexity with speed and automation.

Two different concrete examples will suffice to illustrate the problem of impulsive noise in measurement. In the first a high sample rate representation of a missile trajectory, as measured by radar, is to be analyzed for various purposes. Various sections are seriously contaminated by large impulses. An elaborate analysis can reconstruct the general trajectory shape, and even features like control oscillations and other possible vibrations can be extracted by elaborate techniques. If this is to be done in real time, it is a different problem altogether.

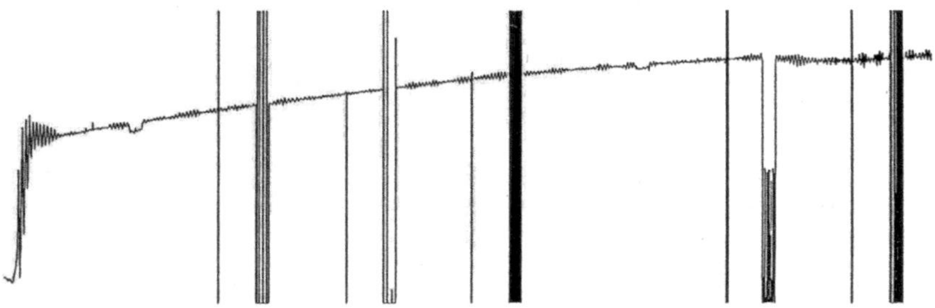

Figure 1. Trajectory data corrupted by impulsive noise.

The second example is data from a speedometer of a small fast boat on a rough sea. The downward impulses, clearly due to the boat leaping out of the water, are significantly larger than the upward impulses. This bias is dependent on sea state. How is the bias to be removed to give a good speed estimate over a large range of sea states?

Statistics has provided an impressive intellectual framework for containing the uncertainty introduced by unexplained or random perturbation ("noise") as exhibited by deviation from the expected "signals" A philosophically consistent way of treating impulsive noise, in the presence of the usual random noise, is to consider the distribution of the random noise as "heavy tailed". This leads to Robust Statistics, as developed by Huber, Hampel and others.

When economy of effort or speed is required a different approach was advocated by Tukey and others. The idea is to view a sequence x of data as the sum of signals, well-behaved noise from a reasonable distribution and added impulsive noise (arbitrarily large amplitude impulses occurring occasionally). A simple selector, like the running median, is used to remove impulses of prescribed width. This is followed by linear smoothing and/or linear signal extraction. The underlying idea is that a linear process "smears" the energy of impulses over neighboring sequence elements, and makes it more difficult to extract afterwards. It is therefore advan-

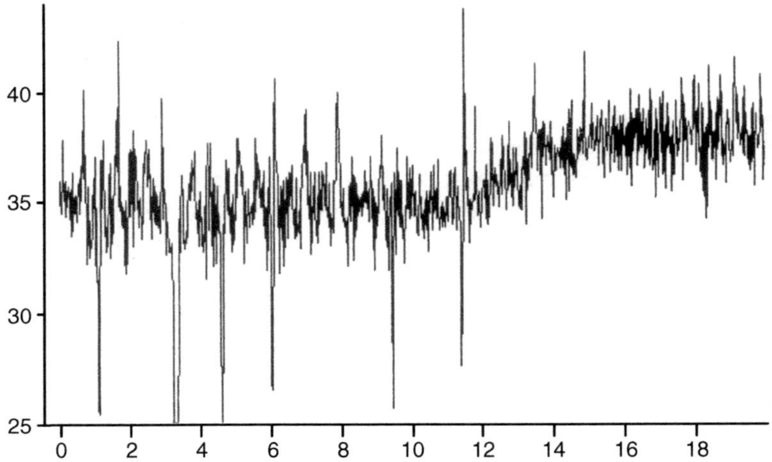

Figure 2. Data from a boat speedometer with impulsive noise.

tageous to remove and replace it, by comparison to immediate neighbors as early as possible, to lessen the damage by the subsequent linear smoothing processes. This is easy to understand and implement and relatively fast and economical, and leads to the subsequent established theory for signal extraction.

These simple nonlinear smoothers (selectors and combinations of them) have over the last two decades received considerable attention, particularly in the engineering literature. Comparisons and analysis have been based mainly on the transfer of distributions if the noise is identically independently distributed random noise. A characterization of "roots", or sequences that remain unchanged by smoothing, did slowly emerge in terms of "edges", "constant regions" and "oscillations", but the "enigmatic behavior" and lack of theory has been almost universally lamented (31). The design and choice of such smoothers has been based mainly on wisdom and experience. Tukey and others popularized smoothers like the three point median (M_1), the three point median applied twice and followed by the five point running median ($M_2M_1^2$), the repeated three point median ($R = M_1^\infty$) and the recursive version (M_1^*) of the three point median. Other rank order selectors are often mentioned, particularly the use of min/max operators in image processing, but the emphasis was on running medians and compositions ("concatenations") of them.

The idea of the median smoother links naturally with that of linear smoothers (filters) in the following way. Instead of projecting a given sequence x onto a chosen subspace ("signals"), a process that may be non-local, it is practical to consider sections of the sequence separately, by passing a "window" over it. Considering an element x_i of x and the neighbors in the window $\{x_{i-n}, \ldots, x_i, \ldots, x_{i+n}\}$, they are used in a projection onto a subspace (constant sequences, linear sequences or sampled higher order polynomials or trigonometric functions). The norm usually used is the $p = 2$ norm, leading to a linear mapping, for instance onto the best constant approximation to $\{x_{i-n}, \ldots, x_i, \ldots, x_{i+n}\}$, from which a smoothed value, (the average of the elements in the window), is substituted for. This is repeated for each i.

Another motivation for this particular norm is usually loosely given as the "Central Limit Theorem". Many decades ago Whitaker and Robinson, in their book "Calculus of Observations", already pointed out the often inappropriate addiction to, or misinterpretation of, the central limit theorem by quoting a witty remark of Poincaré;

"Everybody believes in the exponential law of errors; the experimenters, because they think it can be proved by mathematicians; and the mathematicians, because they believe it has been established by observation."

The Gaussian (normal-) distribution for "noise" on measurement can be motivated by the observation that if a large amount n of numbers, with typical roundoff error from a uniform distribution, are added or subtracted, the result is a spline distribution of order n results. For $n = 6$ this already is almost indistinguishable on a typical graph from a Gaussian. When other operations like division are however also employed, they may result in vastly different error distributions. This problem has been amplified by the vast increase in the number of calculations performed in a modern system. Different methods of adjusting have to be explored.

The convenience of a linear projection associated with the least squares norm is however a good motivation, leading to a simple computational and conceptual procedure. The value x_i in a sequence is replaced by the weighted average of it and its neighbors. (Or the output sequence is the discrete convolute with a fixed local sequence.) Although the linear projection in the window is idempotent, the global smoothing of the sequence, considered as a mapping between input sequence and output sequence, is not.

In the presence of impulsive noise it is natural to replace the least squares projection with an approximation in a more robust norm, the ℓ_1-norm. This leads to the replacing of x_i by median $\{x_{i-n}, \ldots, x_{i+n}\}$, and thus the ("running-") median smoother M_n. Since a median is a good estimator of the average, the median and running average smoothers are similar in effect when the noise is "reasonable", but when the noise is from a "heavy-tailed" distribution the median becomes superior.

It is natural to compare a median and similar smoothers by considering its "linear part" (14), providing a method of comparison. A general perception did persist that other methods of comparison and analysis are required to augment this comparison, but the concept of eigenanalysis of nonlinear operators, which is natural for linear operators, reeks of folly. There may however be considerable merit in it. Even a clear definition of signal, in the context of median smoothing, has generally been avoided and only emerged slowly. Once a clear characterization of "signals", or the significant "roots" of median smoothers, emerged, it became natural to consider the problem of "best approximation" onto signals (4). Although this leads to a computationally unappetizing process, there is merit in the insight obtained and a surprising twist in interpretation.

A consistent framework for analysis and comparison of smoothers has been lacking. For linear filters the vectorspace framework has been useful, leading naturally to eigenanalysis and the associated analysis in the "frequency domain".

A natural alternative framework seems to be the semi-group generated by the unsymmetric minimum and maximum selectors, and the natural order on this semi-group. Not only does this lead to computationally convenient smoothers, almost naturally vectorisable, but it leads to a coherent conceptual framework for analysis and comparison of other rank order smoothers and their composition. The so-called $LULU$-structure has emerged as worth aiming for, also in dimensions higher than 1. The $LULU$-operators are actually particular cases of Morphological Filters, as developed by Serra (29) and others. This was discovered by the author in the mid-1990s, and independently by Maragos and Schafer (15), who informed me shortly afterwards. There is however a strong belief that the specific case has particular value. It is an interesting quirk of logic that that slight relaxation of one of the smoother axioms of Mallows leads to the possibility of adding additional useful and natural axioms, that can be satisfied by practical and consistent smoothers.

The development of the so called $LULU$-theory in the following chapters is based on initial practical work on real problems, for which the author had not been educationally prepared, and subsequent occasional publications on the subject (19)–(25). It introduces the ideas and methods of proof developed previously, but deviates where hindsight suggests a development that is simpler, clearer and more suggestive in indicating search directions in the many facets not yet investigated. It attempts to introduce scientists and mathematicians with a general background into a complementary method of analysis of measurements. As Collatz (10) pointed out long ago, the concepts of order, inequalities and lattices are perhaps the more fundamental.

The final result presented is a novel Multiresolution Analysis (MRA), complementary to that of the discrete wavelet (MRA), which decomposes a sequence into subsequences of linearly decreasing resolution. It does so in a very consistent fashion and explains the accepted good performance of Median Transforms and provides many pleasant and unexpected surprises.

To prove a strong consistency, very important attributes of an operator, like Full Trend Preservation, Neighbor Trend Preservation and Co-idempotence are introduced, and demonstrated to be useful and natural. A similarly important measure of smoothness, Local Monotonicity is introduced, motivated and used extensively, and shown to link with the excepted natural measure of smoothness; Total Variation.

Beauty, simplicity and elegance of poetic metaphors have always been useful when exploring in science and engineering, as in chess. The author is convinced, from his own experience, that the $LULU$-theory is the natural self-contained route into the wonderful field of Mathematical Morphology for all scientists and engineers educated in the usual type of undergraduate curriculum. The hope is that it becomes a well used and useful scenic highway.

1. Operators on Sequences

Tu mir keine Wunder zulieb
Gib deinen Gesetzen zurecht
die von Geschlecht zu Geschlecht
sichtbarer sind.

Rainer Maria Rilke

For the study of sequences it is convenient to use a framework that yields creative insight. The vectorspace framework is convenient and it utilizes natural geometric insight. Let X be the set of bi-infinite sequences $x = \langle x_i \rangle$ of real numbers. As usual, the definition of addition and scalar multiplication is given by:

Definition. $x \oplus y = \langle x_i + y_i \rangle$, for $x = \langle x_i \rangle$ and $y = \langle y_i \rangle$ in X.
$\alpha \odot y = \langle \alpha x_i \rangle$, for $y \in X$ and α a real number.

With these definitions X becomes a vectorspace, and the simpler notation $x + y$ and αx can be used when no misunderstanding should arise. Depending on need and utility, different metrics, norms and inner products can be chosen to make X a metric-, normed- and inner product space respectively. The following norms usually suffice, particularly in the cases of $p = 1, 2$ and ∞.

Definition.
$$||x||_p = \left(\sum_{-\infty}^{\infty} |x_i|^p \right)^{\frac{1}{p}}, \text{ for } p = 1, 2, \ldots,$$
$$\text{and } ||x||_\infty = \sup_i \{|x_i|\}, \text{ for } p = \infty.$$

Definition. $(x, y) = \left(\sum_{-\infty}^{\infty} x_i^2 \right)^{\frac{1}{2}}$ is the usual inner product.

Remarks

(a) Although sequences may generally in practice be finite, zeros can be added to make them bi-infinite and bounded in any of the usual norms.
(b) The topologies induced by the above norms usually suffice.
(c) Unless otherwise stated we shall assume that all sequences are in ℓ_1, or $||x||_1 < \infty$.

Collatz considered orders to be even more basic and useful in numerical mathematics, and this turns out to be justified in the theory of nonlinear smoothers. As is usual an order relation is defined using the usual definition of a relation as an ordered pair of elements from a set.

Definition. Let A be a set and R a relation in A. R is a partial order (partial order relation) if the following are satisfied.

(a) R is reflexive: $(a,a) \in R, \quad \forall a \in A$.
(b) R is anti-symmetric: $(a,b) \in R$ and $(b,a) \in R$ implies $a = b$.
(c) R is transitive: $(a,b) \in R$ and $(b,c) \in R$ imply $(a,c) \in R$.

The usual notation $a \leq b$ for $(a,b) \in R$ and $a < b$ if $(a,b) \in R$ and $a \neq b$ can be introduced. The word "partial" is introduced to suggest that not all pairs in A need be in R. If this is so, it becomes a total order, like the usual order on the real numbers, which is chosen as the point of departure here.

Definition. For $x, y \in X$, $x \leq y$ if and only if $x_i \leq y_i \ \forall i$.

It is standard theory that the above definition results in a partial order on X, but not in a total order. This order on X induces a natural partial (not total) order on the set of operators on X (mappings from X to X) in the following way.

Definition. If A and B are operators on X, then $A \leq B$ if and only if $Ax \leq Bx$, $\forall x \in X$.

It is useful to introduce briefly the usual notation and operators on X.

Definition. Let $F(X)$ be the set of operators on X. Then:

(a) $(A+B)x = Ax + Bx, \quad \forall x \in X$.
(b) $Ix = x, \quad \forall x \in X$.
(c) $Ox = 0, \quad \forall x \in X$ and 0 the null sequence.
(d) $(\alpha A)x = \alpha(Ax), \quad \forall x \in X$.
(e) $(AB)x = A(Bx), \quad \forall x \in X$.
(f) $(Ex)_i = x_{i+1}, \quad$ for each i and $x \in X$.
(g) $Nx = -x, \quad \forall x \in X$.

With the above definition it is standard theory that the addition defines a commutative group of operators, and with the scalar multiplication also a vectorspace for the linear ones. The definition of multiplication (composition) results in a semigroup with identity I, which is non-commutative. Combining all the definitions results in a near-ring of operators. Important to note is that the operators are generally right-distributive, since, by definition of addition, $(A+B)C = AC + BC$ for all $A, B, C \in F(X)$. Unless the operators are all linear, the operators are not generally left-distributive. Although composition is not commutative, non-negative integer powers of a single operator can be shown to commute. Thus, the following can be introduced in a prevalent way.

Definition. $A^0 = I$ and $A^{n+1} = AA^n$, $n = 0, 1, 2, \ldots$.

Theorem 1.1. $A^m A^n = A^n A^m$ for n, m non-negative integers.

Proof. The proof is standard, utilizing induction. □

Definition. An operator S is syntone if $x \geq y \Rightarrow Sx \geq Sy$.

Theorem 1.2. If A and B are syntone operators, then AB is also syntone.

Proof. Let $x \geq y$. Then $Bx \geq By$ and $ABx = A(Bx) \geq A(By) \geq ABy$. □

Theorem 1.3. Let A, B and D be operators on X and A syntone. Then $B \leq D \Rightarrow AB \leq AD$.

Proof. $ABx = A(Bx) \leq A(Dx) = ADx$, for each $x \in X$. □

As point of departure the heuristic motivation for smoothing, namely separating some "signal" from a sequence where the signal is "hidden" (or contaminated) by "noise", should be called to mind. Some simple criteria for design and comparison are appropriate.

Effectiveness	For each x, Px should be a signal and $(I - P)x$ noise.
Consistency	Signals should be preserved and noise mapped onto 0.
Stability	Small input perturbations should not distort output excessively.
Efficiency	The computations should be economical.

Some of these have been formalized implicitly by sets of axioms, such as the axioms of Mallows. Serra suggests idempotence and syntoneness as axioms (29) for "morphological filters". For the purpose at hand, the axioms of Mallows can be taken as the point of departure, weakened slightly to yield the principal smoother axioms. For an elite type, idempotence and co-idempotence can be added to yield the axioms of separators.

For consistency it is reasonable and necessary to demand that the concepts of signal and noise should be translation independent on both axes. This leads to the demands that $P(x+c) = P(x)+c$ and $P(E^j x) = E^j(Px))$, which essentially yield the first two axioms of Mallows. It is also reasonable to demand scale independence. This requires that $P(\alpha x) = \alpha P(x)$, for $\alpha \geq 0$. (Mallows omits the restriction to $\alpha \geq 0$, which seems too lenient). If 0 is a signal it is also noise, since $(I - P)0 = 0$. This yields the following axioms.

Smoother Axioms. An operator P on X is a smoother if:

1. $PE = EP$.
2. $P(x + c) = P(x) + c$, for each $x, c \in X$ such that c is a constant sequence.
3. $P(\alpha x) = \alpha P(x)$, for each $x \in X$ and scalar $\alpha \geq 0$.

Separator Axioms. A smoother P is a separator if it also satisfies the additional axioms:

4. $P^2 = P$ (Idempotence).
5. $(I - P)^2 = I - P$ (Co-idempotence).

The separator axioms can be heuristically argued by a simple analogy to be reasonable demands.

Suppose P is a separator that separates milk x into two components; Px, the curd, and $(I-P)x$, the whey. If P is the only instrument at our disposal, what simple tests could be performed to test the performance of P and what procedure could be followed to improve the performance? A simple test would be to take the curd, and pass it through the separator again. Alternatively, or additionally, we could pass the whey through the separator again. This is schematically illustrated in the following diagram.

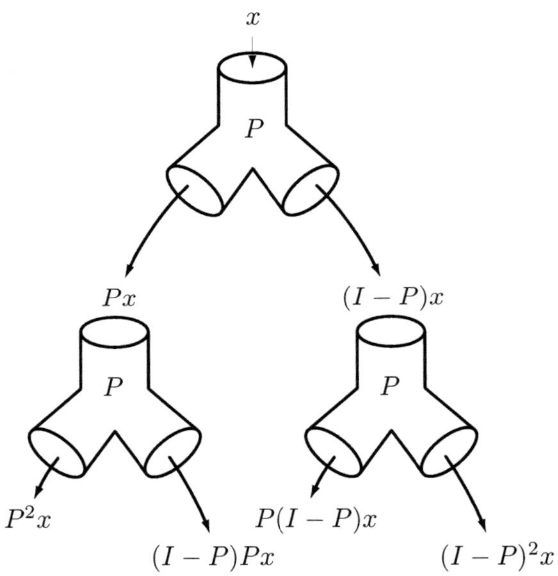

Figure 1.1. Diagram of a two-stage separator cascade.

Interpretations of what happens include the following, where the curd is arbitrarily chosen to be the signal and the whey the noise. (This arbitrary choice is significant.)

1. If $P^2x = Px$, for each sample x, then the separator can be considered signal-consistent.
2. If $(I-P)^2x = (I-P)x$, for each sample x, the separator can be called co-idempotent and considered noise-consistent.
3. A perfect consistent separator is both idempotent and co-idempotent.
4. If a separator P is not idempotent it can be considered a deficient noise-extractor. A better separation could then be achieved by passing the signal through P again. Then P^2x can be considered the signal and $(I-P)Px + (I-P)x$ the noise.

5. If a separator P is not co-idempotent, then it can be considered a deficient signal-extractor. A better separation could be achieved by passing $(I-P)x$ through P again. Then $(I-P)^2x$ can be considered the noise and $Px+P(I-P)x$ the signal.
6. If P is neither idempotent nor co-idempotent, then a better separation can be achieved by passing Px through P again and also passing $(I-P)x$ through P again, and considering $P^2x+P(I-P)x$ as the signal yield and $(I-P)Px+(I-P)^2x$ as the noise yield.

The above interpretations are self-evident but often overlooked. One of the reasons for this is that for projections the separation is consistent, since then the separator is idempotent and co-idempotent. Even for nonlinear operators like best approximations from subspaces, the operator is idempotent and co-idempotent, as 0 is in each subspace. The following is a useful observation.

Lemma. P is co-idempotent if and only if $P(I-P)=0$, where 0 is the operator mapping all sequences onto the zero sequence.

Proof. $I-P=(I-P)^2=(I-P)(I-P)=I-P-P(I-P)$.
This is true if and only if $P(I-P)=0$. □

The lemma confirms directly that for linear operators, idempotence and co-idempotence coincide. This may be the reason why co-idempotence has been generally overlooked as a useful concept. The following simple examples, communicated privately by van der Walt and le Riche respectively, demonstrate that not all nonlinear idempotent operators are co-idempotent and that some are. It therefore seems to make sense to distinguish between them.

Example 1. Let $Px=|x|=\{|x_i|\}$. Clearly $P^2=P$, since $P^2x=P|x|=\{||x_i||\}=\{|x_i|\}=Px$. $(I-P)x=\{x_i-|x_i|\}$ and $(I-P)^2x=2(I-P)x$. Therefore P is not co-idempotent.

Example 2. Let $Px=\{y_i\}$ such that

$$y_i=\begin{cases} x_i & \text{if } x_i\geq 0 \\ 0 & \text{if } x\leq 0 \end{cases}.$$

Then $P^2x=Px$ and $(I-P)x=\{t_i\}$ such that

$$t_i=\begin{cases} x_i & \text{if } x_i< 0 \\ 0 & \text{if } x_i\geq 0 \end{cases}.$$

But then $P(I-P)x=0$ so that $(I-P)^2=I-P$ and P is idempotent. P is therefore a separator. (It is also clear that there are only two eigenvalues; 1 and 0.)

Important classes of basic smoothers are order selectors (rank order selectors), order based selectors and selectors. An n-window is successively centered at each index i to yield the support $\{i-n, i-n+1, \ldots, i, \ldots, i+n\}$ so that the set of values $W_i = \{x_{i-n}, \ldots, x_i, \ldots, x_{i+n}\}$ is selected from the sequence x.

Definition. An order selector (rank order selector) S maps the sequence x onto Sx such that $(Sx)_i = x_j \in W_i$, and such that x_j is a given rank in the order on W_i. A rank based selector uses rank and other information to select an element from W_i and a selector need not use the rank or order in W_i.

To prove that order selectors are smoothers is simple, but additionally they are syntone.

Theorem 1.4. Rank Order selectors are syntone.

Proof. Let S be an order selector. Then $(Sx)_i$ is a given rank, say the kth largest element of $\{x_{i-n}, \ldots, x_i, \ldots, x_{i+n}\}$.

If $y \geq x$ then the kth largest element of $\{y_{i-n}, \ldots, y_i, \ldots, y_{i+n}\}$ is not smaller than $(Sx)_i$. Thus $(Sx)_i \leq (Sy)_i$.

But this is true for each index i, and therefore $Sx \leq Sy$, by the definition of order on sequences. □

All selectors, and also nonrecursive digital (linear-) filters, are local operators in the sense that the value of the image of x at the index i is independent of the values of x_j outside the n-window centered at i. Features of a sequence far away do not influence the image locally.

Theorem 1.5: The set of smoothers form a semigroup with composition and the usual identity.

Proof. I is a smoother, as it satisfies all the axioms. Let A and B be smoothers. Then clearly $(AB)(0) = A(B(0)) = A(0) = 0$, by axiom 1. Also $AB(x+c) = A(B(x)+c) = A(B(x)) + c$, by axiom 2. If E is the shift operator then $(AB)E = A(EB) = (AE)B = (EA)B = E(AB)$ by Axiom 3. □

It is clear that the composition of two idempotent operators is again idempotent if they commute. However, compositions of idempotent operators can be idempotent without commuting. (The operator $LnUn$, as defined later, is idempotent, but $LnUn \neq UnLn$.)

Theorem 1.6. Let A, B be smoothers and $\alpha + \beta = 1$. Then $\alpha A + \beta B$ is a smoother.

Proof.
$$\begin{aligned}
(\alpha A + \beta B)(x+c) &= \alpha A(x+c) + \beta B(x+c) \\
&= \alpha(A(x)+c) + \beta(B(x)+c) \\
&= \alpha A(x) + \beta B(x) + c = (\alpha A + \beta B)(x) + c, \\
(\alpha A + \beta B)E &= \alpha AE + \beta BE = \alpha EA + \beta EB \\
&= E(\alpha A) + E(\beta B) = E(\alpha A + \beta B), \\
(\alpha A + \beta B)(\gamma x) &= \alpha A(\gamma x) + \beta B(\gamma x) = \alpha\gamma A(x) + \beta\gamma B(x), \text{ if } \gamma \geq 0 \\
&= \gamma(\alpha A + \beta B)(x).
\end{aligned}$$
□

Definition. An operator A is linear if $A(\alpha x + \beta y) = \alpha A(x) + \beta A(y)$ for all $x, y \in X$, $\alpha \in \mathbb{R}$.

The linear operators form a ring of operators since they are right-distributive, and are a very important class of operators. Linear (digital-) filters are a special class of smoothers that have extensive applications in many fields. There is a well-developed theory for analysis, design and comparison (11). The theory is based on eigenvalue analysis and statistical analysis, and is associated with approximation, and closely linked to Fourier Analysis, Wavelet Analysis etc. Digital filters are operators that are linear combinations of the form $Fx_i = \sum_j \alpha_j E^j x_{i-j}$. In this sense the shift operators are a basis for the vectorspace of linear operators on sequences. For nonlinear operators an alternative is to consider compositions of "basic" smoothers in a semi-group structure.

2. Basic Rank Selectors, Pulses and Impulses

Zwei mal zwei gleich vier ist Wahrheit.
Schade, daß sie leicht und leer ist.
Denn ich wollte lieber Klarheit.
Über das, was voll und schwer ist.

Wilhelm Busch

A general perception has been prevalent that medians are the "basic" smoothers required to generate other specifically designed composite smoothers.

The smoothers M_n, M_n^k and $R = \lim_{k\to\infty} M_n^k$ have been popularized by Tukey. Mallows and others have extensively investigated them for statistical properties, and by analogy with digital filters, studied the "linear" component of such smoothers. Other popular smoothers derived from the "basic" smoothers M_n are the smoothers like M_2, M_1^2, $2M_n - M_n^2$ and the recursive versions of M_n. Winsorisers are based on the "basic" rank selectors that use quartiles of the elements in a window.

Extensive experience in linear smoothing and statistics may have obscured the possibility of obtaining insight by breaking up the conceptual "atoms" into "positive" and "negative" constituents, that are more basic selectors. Although they do not make up the "atoms" exactly, they suggest a possibility, using order and syntoneness in the semi-group structure, to compare and analyze many smoothers, including the popular rank order smoothers and composition. They also suggest search directions for the construction of smoothers that have many optimal, or near optimal properties.

Considering the medians as "basic" smoother may have had the result of an almost universal exceptance that analysis of nonlinear smoothers is difficult, if not impossible. Considering the extreme selectors, the minimum ("erosions") and the maximum ("dilations"), as "basic" constituents of composite smoothers for the removal of impulsive noise may well be the better alternative. The use of compositions ("openings") and ("closings") of these "basic" selectors has been well known in image processing (29), but a comprehensive theory for understanding the "enigmatic" behavior of nonlinear smoothers comes from their compositions. They can be shown to allow a slightly stricter set of axioms than had been accepted, in the form of those of Mallows.

The following development is not the historical path of discovery, but provides some additional insight, and suggests efficient recursive evaluation.

Definition. Let the operators $\bigvee$ and $\bigwedge$ that map the sequence x onto y be defined by

$$\bigvee: \qquad y = \bigvee x = \langle y_i = \max\{x_i, x_{i+1}\}\rangle,$$
$$\bigwedge: \qquad y = \bigwedge x = \langle y_i = \min\{x_{i-1}, x_i\}\rangle.$$

Theorem 2.1. $\bigvee, \bigwedge$ and I are syntone.

Proof. Let $x > y$. Then, for each index i it follows that,

$$(\bigvee x)_i = \max\{x_i, x_{i+1}\} \geq \max\{y_i, y_{i+1}\} = (\bigvee y)_i.$$

Similarly it follows that $\bigwedge$ is syntone, and the proof for I is trivial. □

Although classified as smoother, since it is a selector, $\bigvee$ or $(\bigwedge)$ does not comply with our intuition as far as smoothers are concerned. This is easily demonstrated on a constant sequence with a single superimposed impulse, as is given by the sequence $x = \langle x_i = \delta_{ij}\rangle$ when $\bigvee x = \langle(\bigvee x)_i = \delta_{i,j} + \delta_{i,j+1}\rangle$. The pulse has been spread and needs "deconvolution" with $\bigwedge$ so that $\bigwedge\bigvee x = x$. This seems a futile exercise, but a sequence with a downward pulse, like $-x$ will have it removed by $\bigvee$, and the operator $\bigwedge$ cannot recreate it, so that $\bigwedge\bigvee(-x) = (-x)$. To remove both upward and downward isolated single impulses we need the compositions $\bigwedge\bigvee$ and $\bigvee\bigwedge$, which are not the same, as the above example demonstrates. The following result is somewhat more subtle than the obvious inequality $\bigwedge \leq I \leq \bigvee$.

Theorem 2.2. $\bigvee\bigwedge \leq I \leq \bigwedge\bigvee$.

Proof. $(\bigvee\bigwedge x)_i = \max\{\bigwedge x_i, \bigwedge x_{i+1}\}$

$$= \max\{\min\{x_i, x_{i-1}\}, \min\{x_i, x_{i+1}\}\} \leq \max\{x_i, x_i\} = (Ix)_i.$$

The rest follows similarly. □

Theorem 2.3. For each non-negative integer n,

$$\overset{n+1}{\bigvee}\overset{n+1}{\bigwedge} \leq \overset{n}{\bigvee}\overset{n}{\bigwedge} \quad \text{and} \quad \overset{n+1}{\bigwedge}\overset{n+1}{\bigvee} \geq \overset{n}{\bigwedge}\overset{n}{\bigvee}.$$

Proof.

$$\overset{n+1}{\bigvee}\overset{n+1}{\bigwedge} = \overset{n}{\bigvee}(\bigvee\bigwedge)\overset{n}{\bigwedge} \quad \text{(associativity)}.$$
$$\leq \overset{n}{\bigvee} I \overset{n}{\bigwedge} \quad \text{(syntoneness of } \bigvee^n\text{)}.$$

A similar proof follows for the rest. □

Theorem 2.4. (The first swallowing theorem). For each $n \leq m$ it follows that

$$\overset{n}{\bigwedge}\overset{n}{\bigvee}\overset{m}{\bigwedge}\overset{m}{\bigvee} = \overset{m}{\bigwedge}\overset{m}{\bigvee} \quad \text{and} \quad \overset{n}{\bigvee}\overset{n}{\bigwedge}\overset{m}{\bigvee}\overset{m}{\bigwedge} = \overset{m}{\bigvee}\overset{m}{\bigwedge}.$$

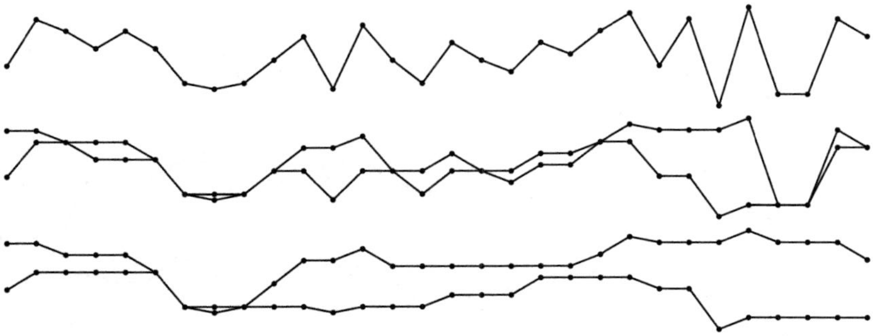

Figure 2.1. Graphs of a random sequence $x, \bigvee\bigwedge x, \bigwedge\bigvee x, \bigvee^2\bigwedge^2 x$ and $\bigwedge^2\bigvee^2 x$.

Proof.
$$\overset{n}{\bigwedge}\Big(\overset{n}{\bigvee}\overset{n}{\bigwedge}\Big)\overset{m-n}{\bigwedge}\overset{m}{\bigvee} \le \overset{n}{\bigwedge} I \overset{m-n}{\bigwedge}\overset{m}{\bigvee} \le \overset{m}{\bigwedge}\overset{m}{\bigvee}.$$

Furthermore
$$\left(\overset{n}{\bigwedge}\overset{n}{\bigvee}\right)\left(\overset{m}{\bigwedge}\overset{m}{\bigvee}\right) \ge I \overset{m}{\bigwedge}\overset{m}{\bigvee},$$

so that the first part of the theorem is proved. The rest follows similarly. □

When $n = m$ this can also be called the first idempotence theorem.

Theorem 2.5. For each non-negative integer n,

$$\overset{n}{\bigvee}\overset{n}{\bigwedge} \le \overset{n}{\bigvee}\overset{n}{\bigwedge}\left(\overset{n}{\bigwedge}\overset{n}{\bigvee}\right)\overset{n}{\bigvee}\overset{n}{\bigwedge} \le \overset{n}{\bigvee}\overset{n}{\bigwedge}\left(\overset{n}{\bigwedge}\overset{n}{\bigvee}\right) \quad \text{and}$$

$$\overset{n}{\bigwedge}\overset{n}{\bigvee}\left(\overset{n}{\bigvee}\overset{n}{\bigwedge}\right) \le \overset{n}{\bigwedge}\overset{n}{\bigvee}\left(\overset{n}{\bigvee}\overset{n}{\bigwedge}\right)\overset{n}{\bigwedge}\overset{n}{\bigvee} \le \overset{n}{\bigwedge}\overset{n}{\bigvee}.$$

Proof.
$$\begin{aligned}
\overset{n}{\bigvee}\overset{n}{\bigwedge} &= \overset{n}{\bigvee}\overset{n}{\bigwedge}\left(\overset{n}{\bigvee}\overset{n}{\bigwedge}\right)\overset{n}{\bigvee}\overset{n}{\bigwedge} && \text{(Theorem 2.4)}\\
&\le \overset{n}{\bigvee}\overset{n}{\bigwedge}\left(\overset{n}{\bigwedge}\overset{n}{\bigvee}\right)\overset{n}{\bigvee}\overset{n}{\bigwedge} && \text{(Theorem 2.2, 2.3)}\\
&\le \overset{n}{\bigvee}\overset{n}{\bigwedge}\left(\overset{n}{\bigwedge}\overset{n}{\bigvee}\right)\overset{0}{\bigvee}\overset{0}{\bigwedge} && \text{(Theorem 2.3)}\\
&= \overset{n}{\bigvee}\overset{n}{\bigwedge}\left(\overset{n}{\bigwedge}\overset{n}{\bigvee}\right).
\end{aligned}$$
□

A similar argument proves the rest.

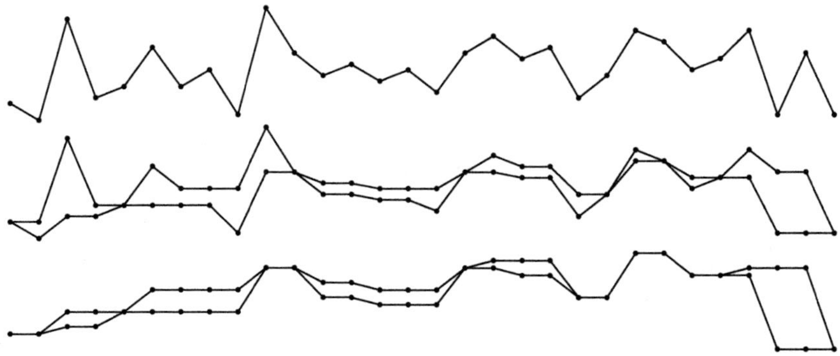

Figure 2.2. Graphs of $x, \bigvee\bigwedge x, \bigwedge\bigvee x, \bigwedge\bigvee\bigvee\bigwedge x$ and $\bigvee\bigwedge\bigwedge\bigvee x$.

Numerical experiments indicate that $\bigwedge\bigvee\bigvee\bigwedge x \leq \bigvee\bigwedge\bigwedge\bigvee x$, and although this is true, it cannot be proved from the semi-group structure and the order alone. However the following three theorems follow algebraically.

Theorem 2.6. (The second idempotence theorem). For each non-negative integer n;

$$\left[\overset{n}{\bigvee}\overset{n}{\bigwedge}\left(\overset{n}{\bigwedge}\overset{n}{\bigvee}\right)\right]^2 = \overset{n}{\bigvee}\overset{n}{\bigwedge}\left(\overset{n}{\bigwedge}\overset{n}{\bigvee}\right) \quad \text{and} \quad \left[\overset{n}{\bigwedge}\overset{n}{\bigvee}\left(\overset{n}{\bigvee}\overset{n}{\bigwedge}\right)\right]^2 = \overset{n}{\bigwedge}\overset{n}{\bigvee}\left(\overset{n}{\bigvee}\overset{n}{\bigwedge}\right).$$

Proof.

$$\begin{aligned}
\overset{n}{\bigvee}\overset{n}{\bigwedge}\left(\overset{n}{\bigwedge}\overset{n}{\bigvee}\right) &= \overset{n}{\bigvee}\overset{n}{\bigwedge}\left(\overset{n}{\bigwedge}\overset{n}{\bigvee}\overset{n}{\bigwedge}\overset{n}{\bigvee}\right) && \text{(Theorem 2.4)} \\
&= \left(\overset{n}{\bigvee}\overset{n}{\bigwedge}\overset{n}{\bigwedge}\overset{n}{\bigvee}\right)\overset{n}{\bigwedge}\overset{n}{\bigvee} \\
&\geq \left(\overset{n}{\bigvee}\overset{n}{\bigwedge}\overset{n}{\bigwedge}\overset{n}{\bigvee}\right)\overset{n}{\bigvee}\overset{n}{\bigwedge}\overset{n}{\bigwedge}\overset{n}{\bigvee} && \text{(Theorem 2.2, 2.3)} \\
&= \overset{n}{\bigvee}\overset{n}{\bigwedge}\left(\overset{n}{\bigwedge}\overset{n}{\bigvee}\right)\overset{n}{\bigvee}\overset{n}{\bigwedge}\overset{n}{\bigwedge}\overset{n}{\bigvee} \\
&\geq \overset{n}{\bigvee}\overset{n}{\bigwedge} \; I \; \overset{n}{\bigvee}\overset{n}{\bigwedge}\overset{n}{\bigwedge}\overset{n}{\bigvee} \\
&= \overset{n}{\bigvee}\overset{n}{\bigwedge}\overset{n}{\bigwedge}\overset{n}{\bigvee}. && \text{(Theorem 2.4)}
\end{aligned}$$

The rest follows similarly. □

Theorem 2.7. $$\overset{n}{\bigvee}\overset{n}{\bigwedge}\overset{n}{\bigwedge}\overset{n}{\bigvee}\overset{n}{\bigvee}\overset{n}{\bigwedge} \leq \overset{n}{\bigwedge}\overset{n}{\bigvee}\overset{n}{\bigvee}\overset{n}{\bigwedge}\overset{n}{\bigwedge}\overset{n}{\bigvee}.$$

Proof.

$$\overset{n}{\bigvee}\overset{n}{\bigwedge}\overset{n}{\bigwedge}\overset{n}{\bigvee}\overset{n}{\bigvee}\overset{n}{\bigwedge} \leq I\overset{n}{\bigwedge}\overset{n}{\bigvee}\overset{n}{\bigvee}\overset{n}{\bigwedge} \leq \overset{n}{\bigwedge}\overset{n}{\bigvee}\overset{n}{\bigvee}\overset{n}{\bigwedge}I$$
$$\leq \overset{n}{\bigwedge}\overset{n}{\bigvee}\overset{n}{\bigvee}\overset{n}{\bigwedge}\overset{n}{\bigwedge}\overset{n}{\bigvee}.$$

□

Theorem 2.8. (The third idempotence theorem). For each non-negative integer n;

$$\left[\overset{n}{\bigvee}\overset{n}{\bigwedge}\overset{n}{\bigwedge}\overset{n}{\bigvee}\overset{n}{\bigvee}\overset{n}{\bigwedge}\right]^2 = \overset{n}{\bigvee}\overset{n}{\bigwedge}\overset{n}{\bigwedge}\overset{n}{\bigvee}\overset{n}{\bigvee}\overset{n}{\bigwedge} \quad \text{and}$$
$$\left[\overset{n}{\bigwedge}\overset{n}{\bigvee}\overset{n}{\bigvee}\overset{n}{\bigwedge}\overset{n}{\bigwedge}\overset{n}{\bigvee}\right]^2 = \overset{n}{\bigwedge}\overset{n}{\bigvee}\overset{n}{\bigvee}\overset{n}{\bigwedge}\overset{n}{\bigwedge}\overset{n}{\bigvee}.$$

Proof.

$$\overset{n}{\bigvee}\overset{n}{\bigwedge}\overset{n}{\bigwedge}\overset{n}{\bigvee}\left(\overset{n}{\bigwedge}\overset{n}{\bigvee}\overset{n}{\bigvee}\overset{n}{\bigwedge}\right)\overset{n}{\bigwedge}\overset{n}{\bigvee}\overset{n}{\bigvee}\overset{n}{\bigwedge}$$
$$= \overset{n}{\bigvee}\overset{n}{\bigwedge}\overset{n}{\bigwedge}\overset{n}{\bigvee}\left(\overset{n}{\bigwedge}\overset{n}{\bigvee}\right)\overset{n}{\bigwedge}\overset{n}{\bigvee}\overset{n}{\bigvee}\overset{n}{\bigwedge} \qquad \text{(Theorem 2.4)}$$
$$= \overset{n}{\bigvee}\overset{n}{\bigwedge}\left(\overset{n}{\bigwedge}\overset{n}{\bigvee}\overset{n}{\bigwedge}\overset{n}{\bigvee}\right)\left(\overset{n}{\bigwedge}\overset{n}{\bigvee}\overset{n}{\bigvee}\overset{n}{\bigwedge}\right)$$
$$= \overset{n}{\bigvee}\overset{n}{\bigwedge}\left(\overset{n}{\bigwedge}\overset{n}{\bigvee}\overset{n}{\bigwedge}\overset{n}{\bigvee}\right). \qquad \text{(Theorem 2.6)}$$

The rest follows similarly. □

With the suggestive notation

$$L_n = \overset{n}{\bigvee}\overset{n}{\bigwedge}, \quad U_n = \overset{n}{\bigwedge}\overset{n}{\bigvee},$$

it follows that;

1. $L_{n+1} \leq L_n \leq L_0 = I = U_0 \leq U_n \leq U_{n+1}$. (Theorem 2.2, 2.3)
2. ($L_n L_m = L_m$ and $U_n U_m = U_m$) for $m \geq n$. (Theorem 2.4)
3. $L_n \leq L_n U_n L_n \leq L_n U_n$ and $U_n L_n \leq U_n L_n U_n \leq U_n$. (Theorem 2.5)
4. $(L_n U_n)^2 = L_n U_n$ and $(U_n L_n)^2 = U_n L_n$ (Theorem 2.6)
5. $L_n U_n L_n \leq U_n L_n U_n$. (Theorem 2.7)
6. $(L_n U_n L_n)^2 = L_n U_n L_n$ and $(U_n L_n U_n)^2 = U_n L_n U_n$.

The following theorem is very useful.

Theorem 2.9. For each integer $n \geq 0$ the following are equivalent.

(a) $LnUn \geq UnLn$.

(b) $LnUnLn = UnLn$.

(c) $UnLnUn = LnUn$.

Proof. Let (a) be true. Then $Ln(UnLn) \leq IUnLn = UnLn$, since $Ln \leq I$. But $(LnUn)Ln \geq (UnLn)Ln = Un(LnLn) = UnLn$, since $LnUn \geq UnLn$ and $Ln^2 = Ln$. Therefore (a) implies (b).

Suppose that $LnUnLn = UnLn$. Then

$$\begin{aligned}(UnLn)Un &= (LnUnLn)Un = (LnUn)(LnUn), \\ &= LnUn,\end{aligned}$$

by the idempotence of $LnUn$. Therefore (b) $\Rightarrow$ (c).

Suppose $UnLnUn = LnUn$. Then

$$LnUn = (UnLn)Un \geq UnLnI,$$

since $UnLn$ is syntone. Therefore (c) $\Rightarrow$ (a).

This completes the proof of the theorem. □

A useful insight obtained is that, for each positive integer n, the operators $L = Ln$ and $U = Un$ form a semi-group that is completely ordered if the inequality $LU \geq UL$ can be proved. In such a case the ordered semi-group shall be called an $LULU$-structure, and its complete multiplication table is;

	L	U	UL	LU
L	L	LU	UL	LU
U	UL	U	UL	LU
UL	UL	LU	UL	LU
LU	UL	LU	UL	LU

The order is then completely determined by: $L \leq UL \leq LU \leq U$.

The fourth swallowing theorem $LUL = UL$ and $ULU = LU$ will be proved via the significant inequality $LnUn \geq Mn \geq UnLn$, where Mn is the well-known median smoother.

It should be clear that in the definition of $\bigvee$ and $\bigwedge$ there is an alternative choice for the shift of index in the maximum and minimum. However, when the combinations $Ln = \bigvee^n \bigwedge^n$ and $Un = \bigwedge^n \bigvee^n$ are considered, it is useful to note that they are unsymmetric only in that $Ln(-x) = -Unx$. This can be formalised by

Theorem 2.10. $LnN = NUn$.

Proof.
$$\bigvee^n x_i = \max\{x_i, x_{i+1}, \dots, x_{x+n}\}$$
$$= -\min\{-x_i, -x_{i+1}, \dots, -x_{i+n}\} \quad \text{and}$$
$$\bigwedge^n x_i = \min\{x_{i-n}, \dots, x_i\} = -\max\{-x_{i-n}, \dots, -x_i\}.$$

But
$$LnNx = Ln(-x) = \left\langle \bigvee^n \left(\bigwedge^n (-x) \right)_i \right\rangle = \left\langle \bigvee^n \left(- \bigvee^n x \right)_i \right\rangle$$
$$= \left\langle - \bigwedge^n \bigvee^n x_i \right\rangle = N \bigwedge^n \bigvee^n x,$$

for each x. Hence $LnN = NUn$. □

Definition. An operator B is the dual of A if $AN = NB$.

It is clear that Ln and Un are duals of each other and that all compositions of operators of these two types are again dual to the operator formed by interchanging Ln and Un, for each n. Any inequality comparing two such compositions changes the inequality around when such duals are interchanged, for example if $LU \leq U$, then
$$NL = UN \geq LUN = LNL = NUL, \quad \text{so that} \quad L \leq UL.$$

It is therefore expedient to speak of dual theorems, and the duality can be used to prove one from the other by simple algebra.

Remarks

1) We note that $\bigvee$ and $\bigwedge$ are not duals of each other but that $\bigvee E^{-1}N = N\bigvee$ or $\bigwedge EN = N\bigvee$, by the definition and the fact that E commutes with $\bigvee, \bigwedge$ and N.
2) Any operator that commutes with N can be considered a dual of itself, and any pair of operators A and B such that $AN = BN$ are duals of each other. Therefore N, I, O, E are all self-dual operators.

At this stage the concept of "signal" and "noise", which have been undefined, can slowly be developed, more or less on historical lines by considering blockpulses.

Definition. A (primitive) n-blockpulse is a blockpulse of length n. It is called upward if it is non-negative and downward if it is non-positive.

Definition. A (primitive) n-pulse is a sequence x such that $x_j = 0$ for $i \notin [j+1, j+n]$ for some integer j.

Theorem 2.11: Let x be an upward n-blockpulse and z a downward n-pulse. For each integer $k \geq 0$ the following are true:

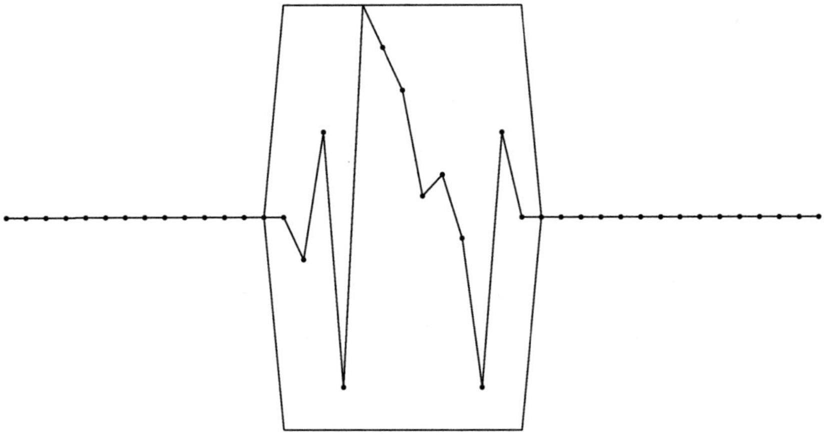

Figure 2.3. An upward blockpulse, a downward blockpulse and a pulse in between.

(a) $\bigwedge^k x$ is an upward blockpulse of length $\max\{0, n-k\}$.

(b) $\bigvee^k x$ is an upward blockpulse of length $n+k$.

(c) $\bigwedge^k z$ is a downward blockpulse of length $n+k$.

(d) $\bigvee^k z$ is a downward blockpulse of length $\max\{0, n-k\}$.

Proof. (a) Induction on k. $\bigwedge^0 = I$ so that $\bigwedge^0 x$ is an upward blockpulse of length n. Assume (a) holds for $k < n$. Then $\exists j$ such that

$$\left(\bigwedge^k x\right)_i = 0 \quad \text{for} \quad i \notin \{j+1, j+2, \ldots, (j+\max\{0, n-k\})\}.$$

But then

$$\left(\bigwedge^{k+1} x\right)_i = \bigwedge\left(\bigwedge^k x\right)_i = \min\left\{\left(\bigwedge^k x\right)_{i-1}, (\bigwedge^k x)_i\right\}$$

for all $i \notin \{j+2, j+3, \ldots, (j+\max\{0, n-k\})\}$. But this is a blockpulse of width $\max\{0, n-(k+1)\}$. If $k = n-1$ then $\bigwedge^{k+1} x = 0$.

The proof for the rest of the theorem follows similarly, from the definition of $\bigvee$ and $\bigwedge$, by induction. □

Clearly upward blockpulses are narrowed by $\bigwedge$ and widened by $\bigvee$ and downward blockpulses are widened by $\bigwedge$ and narrowed by $\bigvee$. Once either has been narrowed to a width of 0, the result is the zero sequence (an elementary blockpulse!).

Theorem 2.12. Let x be an upward blockpulse of length n. Then:

$$\bigvee^m \bigwedge^m x = \begin{cases} 0, & \text{for } m \geq n \\ x, & \text{for } 0 \leq m < n \end{cases}$$

and

$$\bigwedge^m \bigvee^m x = x, \text{ for each } m \geq 0.$$

Proof. Let x be an upward n-blockpulse of amplitude α.
If $m \geq n$, then by Theorem 5.1(a) it follows that $\bigwedge^m x$ is a blockpulse of length $\max\{0, n-m\} = 0$.
If $m < n$, then $\bigwedge^m x$ is a blockpulse of length $n - m$, by Theorem 5.1(a).
By Theorem 5.1(b) $\bigvee^m (\bigwedge^m x)$ is a blockpulse of length $n - m + m = n$.

Clearly the amplitude of a blockpulse remains at the same value α through mapping by $\bigwedge$ and$\bigvee$. By comparing indexes where x and $\bigvee^m \bigwedge^m x$ are zero they are shown to be identical.

Similarly, $\bigvee^m x$ is a blockpulse of length $n+m$ and $\bigwedge^m \bigvee^m x$ is a blockpulse of length n and amplitude α. Comparing indexes where they are zero it is shown that $\bigwedge^m \bigvee^m x = x$. □

Corollary. The Dual theorem of 5.2 holds, since if x is an upward n-blockpulse, Nx is a downward n-blockpulse and $\bigvee^m \bigwedge^m N = N \bigwedge^m \bigvee^m$, since $\bigvee^m \bigwedge^m$ and $\bigwedge^m \bigvee^m$ are dual operators.

Clearly, the last theorem shows that n-blockpulses are annihilated (mapped onto 0) by $\left\{ \begin{array}{c} Lm \\ Um \end{array} \right\}$ provided they are $\left\{ \begin{array}{c} \text{upward} \\ \text{downward} \end{array} \right\}$ respectively if $m \geq n$ and identically preserved if $m < n$. If blockpulses of length $k \leq n$ are considered as noise, then Lm and Um are smoothers for the upward and downward blockpulses respectively, and $LmUm$ would remove both such upward and downward blockpulses, since Um removes the downward blockpulses and preserves the upward pulses, which are then removed by Lm. Similarly $UmLm$ removes both types.

When the blockpulses are wide enough they are perfectly preserved by both composite smoothers $UmLm$ and $LmUm$. These operators clearly meet the primary requirements of smoothers, and are syntone as well. Syntone operators are order preserving. This useful property will be used to prove that they remove a wider class of noise impulses, at least when these are added to a constant sequence. To show this the following definition is required.

Definition. An Impulse (n-impulse)is a sequence x such that $x_i = 0$ for $i \notin \{j+1, \ldots, j+n\}$, for some integer j.

This definition is the same as for an n-pulse, but signifies that it is chosen as impulsive noise.

Theorem 2.13. Let x be an n-impulse. Then $LmUmx = UmLmx = 0$ for $m \geq n$.

Proof. There exist n-blockpulses u and $-u$ such that $-u \leq x \leq u$ with $||u|| < 0$.

$UmLm$ and $LmUm$ are syntone, so that
$UmLm(-u) \leq UmLm(x) \leq LmUm(x) \leq LmUm(u)$.

But $LmUm(u) = 0$ and $-LmUm(u) = UmLm(-u) = 0$,
so that $UmLm(x) = LmUm(x) = 0$. □

Corollary: An n-pulse is removed from any constant sequence by $LmUm$ and $UmLm$, if $m \geq n$.

The operators U and L as defined here meet all the requirements of Morphological Filters (29), and in the terminology used there;

U is "extensive", as $I \leq U$, and

L is "anti-extensive" as $I \geq L$.

Furthermore U is a "closing" as it is "increasing" (syntone), extensive and idempotent and L is an "opening" as it is increasing, anti-extensive and idempotent.

There is an alternative to using compositions of U and L to remove upward and downward impulses from constant sequences. The difference between the pair LU and UL may turn out to be illuminating, but if an unbiased smoother is desired the median is available, since it is an unbiased estimator of the average of the same elements. The following heuristic argument seems sound; since U removes downward pulses and L upward pulses the sum should be twice the signal plus both impulses. Subtracting the signal plus both impulses, namely the sequence x, should leave virtually only the signal.

Definition. $Qn = Un + Ln - I$.

Theorem 2.14. Let x be an n-blockpulse. Then $Qnx = 0$.

Proof. If $x \geq 0$ then $Unx = x$ and $Lnx = 0$ by the above theorem.
Thus $Qx = Ux + Lx - x = x + 0 - x = 0$.
If $x \leq 0$ then $Lnx = x$ and $Unx = 0$, and again $Qx = 0$. □

The operator Q is selfdual since $QN = UN + LN - N = I - U - L = NQ$. Thus it is unbiased, like the median smoother.

Theorem 2.15. Qn is a smoother.

Proof. From Theorem 1.6 it follows that $2Un - I$ and $2Ln - I$ are smoothers.

By the same theorem then, their average is also a smoother
$\frac{1}{2}(2Ln - I) + \frac{1}{2}(2Un - I) = Un + Ln - I$. □

Remarks.

(a) Q_n is not idempotent, since if: $x = \langle(-1)^i\rangle$,
$Unx = 1$ and $Lnx = -1$.
Therefore $Q_nx = -x$, and similarly $Q_n^2x = x$.

(b) Q_n is not necessarily syntone, and not necessarily a selector. However, Q_1 is a syntone selector. It is easy to show that $Q_1 = M_1$ but this identity only holds in $n = 1$.

Clearly Q_n seems as useful and at least as enigmatic in behavior as the median M_n.

At this point some problems also become clearer, since if an n-impulse is added to a general monotone sequence, its removal is not certain. In fact, even its identification is problematic. Consider the case of an impulse $x_i = \delta_{ij}$ added to a monotone sequence $s_j = \alpha i$, for some scalar α. If $\alpha > 1$ then $s_i + x_i$ is monotone increasing, and so is $s_i - x_i$. If α is not known, how is the presence of impulsive noise to be identified at the index j? Even more basic. If a primitive 1-impulse n_j is removed by L_1U_1 and U_1L_1, and another n_k also, what about the sum of two such "noise"-impulses? If $j - k > n$ then they are both removed from constant sequences by L_1U_1 and U_1L_1, but if $j - k = \pm 1$ or $j - k = \pm 2$ they are not.

In both the above examples an ambiguity arises. How is this to be interpreted, particularly with respect to the relatively simple task of extending the definition of signal and noise to larger classes of sequences? Is there any unambiguous definition of noise or signal? If so what are they, and is the characterization illuminating? Furthermore; how much has been achieved in clarifying the enigmatic, erratic behavior of the popular median smoothers? It is clear that the median smoothers, defined by

$$(M_n x)_i = \text{median}\{x_{i-m}, \ldots, x_i, \ldots, x_{i+m}\},$$

remove n-blockpulses from constant sequences as well as L_mU_m and U_mL_m, since the majority of the set $\{x_{i-m}, \ldots, x_{i+m}\}$ are equal to the constant value of the constant sequence. What has been gained? How are smoothers like L_mU_m, U_mL_m and M_m to be compared? And what about the many others that seem competitive, are popular and treated as "basic"?

Is there a sensible, useful definition of signal and noise which separates X into two sets, sharing only the zero sequence? At this stage it seems reasonable that if pulses of width $(n+1)$ are to be considered signals, then at least, impulses of lesser width are noise. The concept will be generalized after this has been done to the concept of signal.

3. *LULU*-Smoothers, Signals and Ambiguity

Wollest mit Freuden
Und wollest mit Leiden
Mich nicht überschütten
Doch in der Mitten
Liegt holdes Bescheiden.
Eduard Mörike

It is clear how the operators $LnUn$ and $UnLn$ treat pulses of sufficient briefness when they are added to constant sequences or to sequences that are constant for a sufficiently large neighborhood of the pulse. Since the operators are non-linear they clearly could be treated differently when added to another sequence, since they can only be detected as pulses with respect to their neighboring values. One way is to characterize the sequences that pass unaltered through $LnUn$ and $UnLn$ and consider these signals, and then to consider the residue $(I - LnUn)x$, and $(I - UnLn)x$ as the noise that is superimposed on the signal. This turns out to be more illuminating than the alternative.

This idea was used with the median smoother Mn, and resulted in a less than simple characterization in terms of "edges", "constant regions" etc. In the case of the *LULU*-operators, this turns out to be simpler.

Theorem 3.1. Let $s < q < t$ and $|t - s| < n + 2$.

$$\begin{aligned} &\text{If} \quad (Lnx)_s < (Lnx)_q, \quad \text{then} \quad (Lnx)_t \geq (Lnx)_q, \quad \text{and} \\ &\text{if} \quad (Unx)_s > (Unx)_q, \quad \text{then} \quad (Unx)_t \leq (Unx)_q. \end{aligned}$$

Proof. Suppose $(Lx)_s < (Lx)_q$ and $(Lx)_t < (Lx)_q$, with L denoting Ln.
Since $(Lx)_s < (Lx)_q$, $(Lx)_q > \min\{x_{j-n}, \ldots, x_j\}$, for each $j \in [s, s+n]$.
Since $(Lx)_t < (Lx)_q$, $(Lx)_q > \min\{x_{j-n}, \ldots, x_i\}$ for each $j \in [t, t+n]$.
Since $|s - t| < n + 2, \min\{x_{j-n}, \ldots, x_j\}, < (Lx)_q$ for $j \in [s, t+n]$, and $(Lx)_q$ is one of these minima, which is a contradiction. Therefore $(Lx)_t \not< (Lx)_q$.

The rest of the theorem follows, since Un is the dual of Ln. □

The above theorem demonstrates that the image $y = Lnx$ of a sequence x cannot have local outliers upwards. Similarly $y = Unx$ so that it cannot have downward outliers. This can be formally proved in the following theorem.

Definition. A sequence x is n-monotone if for each $j, \{x_j, x_{j+1}, \ldots, x_{j+n+1}\}$ is monotone.

Figure 3.1. The output L_1x of a random sequence, showing no points larger than both neighbors.

Theorem 3.2. For each n, $Lnx = Unx$ if and only if x is n-monotone

Proof. Let $Lx = Lnx = Unx = Ux$ and $X = \{x_{i-n}, x_{i-n+1}, \ldots, x_i, x_{i+1}\}$. Since $Lx \leq x \leq Ux$, $\forall x \in X$, it follows that $x = Lx = Ux$.

Let x_q be the first element in X that differs from x_{i-n}. Assume that $x_q > x_{i-n}$. Then $(Lx)_q > (Lx)_{i-n}$ and, by the previous theorem, it follows that $(Lx)_{q+1} \geq (Lx)_q$ and therefore $x_{q+1} \geq x_q$.

By induction it follows that $x_{j+1} \geq x_j$, for each $j \in \{i-n, \ldots, i\}$ showing that the $n+2$ successive numbers are monotone increasing.

If $x_q < x_{i-n}$, a similar argument with U proves that X is monotone decreasing, and thus that $\{x_{i-n}, \ldots, x_{i+1}\}$ is monotone. Since this is true for each index i, the sequence x is n-monotone.

Conversely, consider the set $A = \{x_{i-n}, \ldots, x_{i+n}\}$, with x being n-monotone. If the set is monotone increasing, $\min\{x_j, \ldots, x_{j+n}\} = x_j$, for each $\in [i-n, i-1]$. But then $(Lx)_i = \max\{x_{i-n}, x_{i-n+1}, \ldots, x_i\} = x_i$, and a similar argument shows that if A is increasing then again $(Lx)_i = x_i$.

Suppose therefore that A is not monotone. Then there is a j and two subsets $\{x_{j-1}, \ldots, x_{j+n}\}$ and $\{x_j, \ldots, x_{j+n+1}\}$ of A, such that one is monotone increasing and the other monotone decreasing. The intersection $\{x_j, \ldots, x_{j+n}\}$ must then be constant and contain x_i. Therefore $(Lx)_i$, which is the maximum of the minima of such subsets, one of which is $\min\{x_j, \ldots, x_{j+n}\} = x_i$, is not smaller than x_i.

Since $Lx \leq x$ for any sequence, this means that $(Lx)_i = x_i$.

A similar argument, or duality, shows that $(Ux)_i = x_i$, and since the whole argument is valid for each index i, it follows that $Lx = x = Ux$. □

What is more important for the characterization of the output of the smoothers LU and UL are the following theorems, which characterize the range of both the smoothers in terms of an important concept.

Theorem 3.3. For each n, $LnUnx = x = UnLnx$ if and only if x is n-monotone.

Proof. Let x be n-monotone. Then, by the previous theorem, $Ux = Lx = x$. But then $U(Lx) = Ux = Lx = L(Ux) = x$.

Conversely, let $ULx = LUx = x$. Then $Lx = L(LUx) = LL(Ux) = LUx = x$ and $Ux = U(ULx) = UU(Lx) = ULx = x$.

By the previous theorem, this implies that x is n-monotone. □

Corollary. $UnLnx = x$ if and only if x is n-monotone, and $UnLnx = x$ if and only if $x = LnUnx$.

Proof. $Unx = Un(UnLnx) = UnLnx = x$ and similarly $Lnx = Ln(UnLnx) = UnLnx = x$, by the following theorem's corollary.

It should be remarked that $UnLnx = LnUnx$ does not imply that x is n-monotone. This contrasts with the case when $Unx = Lnx$, where the fact that $Ln \leq I \leq Un$, proves that $Unx = Lnx = x$. But generally $UnLnx = LnUnx$ does not imply that $x = LnUnx = UnLnx$, since generally it is not true that x lies between $UnLnx$ and $LnUnx$.

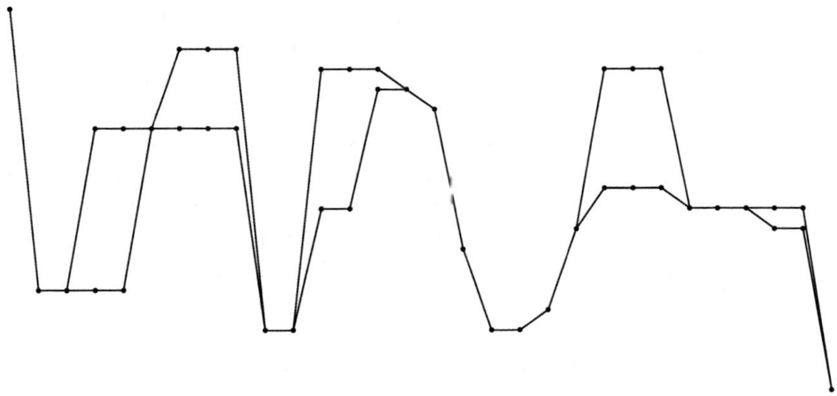

Figure 3.2. ULx and LUx for a random sequence x demonstrating 1-monotone output.

Theorem 3.4. For each integer $n \geq 0$, $UnLn \leq Mn \leq LnUn$.

Proof. Assume $(Mnx)_i > (LnUnx)_i$ for some sequence x and index i.

From the definition of Ln it follows that there are two indexes s and t such that $(Unx)_s, (Unx)_t < (Mnx)_i$, with $i \in [s,t]$ and $[s-t] \leq n$. But then, from the definition of Un, there exist two indexes $j \in [s, s+n]$ and $q \in [t, t+n]$ such that $\max\{x_{j-n}, \ldots, x_j\}, \max\{x_{q-n}, \ldots, x_q\} < (Mnx)_i$.

Consider the union $\{x_{j-n}, \ldots, x_q\} = \{x_{j-n}, \ldots, x_j\} \cup \{x_{q-n}, \ldots, x_q\}$. It contains at least $n+1$ elements from the set $\{x_{i-n}, \ldots, x_{i+n}\}$, which are all smaller than the median, which in turn cannot be larger than more than n of the elements. This contradiction proves the one inequality.

A similar proof yields the other, or alternatively, an argument using the duals of Mn and $LnUn$ concludes the proof of the theorem. □

Corollary. $UnLnUn = LnUn$ and $LnUnLn = UnLn$, by Theorem 2.9.

Theorem 3.5. For each sequence $x \in X$, $UnLnx$ and $LnUnx$ are n-monotone.

Proof. By the above corollary it follows that

$$Un(UnLnx) = (UnUn)(Lnx) = Un(Lnx) = UnLnx \quad \text{and}$$
$$Ln(UnLnx) = (LnUn)(Lnx) = (LnUnLn)x = UnLnx.$$

But then, by Theorem 3.2. $UnLnx$ is n-monotone.
A similar proof follows for $LnUnx$. □

It is illuminating to take stock of what has been achieved. The operators $LnUn$ and $UnLn$ share the range $\mathcal{M}_n$, given by the following definition.

Definition. $\mathcal{M}_n$ is the set of all sequences in x that are n-monotone.

The sequences in $\mathcal{M}_n$ can therefore be considered the signals, as interpreted by both operators, and $LnUn$ and $UnLn$ can be considered to be extensions of the identity operator on the set $\mathcal{M}_n$. The previous two theorems also imply that the sequences are all roots of the operators M_n, M_n^2 and the (nonexistent) operator $R = M^\infty$. Thus all the popular smoothers used to remove the n-impulses share these common "roots", as they were also often called. Some finite powers of M_n do have roots that are not in M_n, but these can be considered an insignificant minority. These roots, not in $\mathcal{M}_n$, can be called spurious roots.

An example is the sequence $x = \langle(-1)^i\rangle$ which is a root of M_2, since clearly $M_2x = x$. (This sequence is precisely a sampling of the Nyquist-frequency , which should hardly be considered as signal.)

It is important to note that the median operator Mn is not equal to either $UnLn$ or $LnUn$.

Example. The sequence $x = \langle(-1)^i\rangle$ yields $Mn(x) = \langle(-1)^{i+n}\rangle$, whereas

$$UnLn(x) = Ln(x) = -1 \quad \text{and} \quad LnUn(x) = Un(x) = 1.$$

The operators are however equal on the very large set $\mathcal{M}_n$, and on a larger set , namely the set of all roots of the operators Un or Ln. This is demonstrated by the following theorem.

Theorem 3.6. For $x = Un(x)$ or $x = Ln(x)$, $Mn(x) = LnUn(x) = UnLn(x)$.

Proof. $LUx = LU(Ux) \geq M(Ux) \geq UL(Ux) = (ULU)x = LUx$. □

Corollary. $MnUn = LnUn$ and $MnLn = UnLn$.

It is important to understand that the operators $MnUn$ and $MnLn$ are not the same as $UnMn$ and $LnMn$ respectively. The latter two are not idempotent operators whereas the above corollary shows that $MnUn$ and $MnLn$ are idempotent.

Theorem 3.7. For each $n \geq 0$ and $j \geq 0$, using the notation $M = M_n$,

$$LnM^j \leq LnM^{j+1} \quad \text{and} \quad UnM^j \geq UnM^{j+1}.$$

Proof.
$$\begin{aligned} LnMn &\leq UnLnMn, && \text{since } I \leq Un, \\ &= LnUnLnMn, && \text{since } UnLn = LnUnLn, \\ &= LnMnLnMn, && \text{since } UnLn = MnLn, \\ &\leq LnMnIMn. \end{aligned}$$

Therefore $LnM^j = (LnMn)M^{j-1} \leq LnM^2nM^{j-1} \leq LnM^{j+1}$.

A similar argument (or an argument with duals) proves the rest of the theorem. □

It is easy to construct particular sequences to show that the inequalities in the theorem cannot be replaced by equalities.

Example. The sequence $x = \langle x_i = \delta_{i,j-1} - \delta_{i,j} + \delta_{i,j+1} \rangle$ containing an alternating triplet of pulses. Since $m_1 x = \langle m_i = \delta_{i,j} \rangle$ and $M_1^2 x = 0$, it is clear that $U_1 M_1 x = M_1 x$ has a unit pulse, whereas $U_1 M_1^2 x = 0$. This example also demonstrates that $U_1 M_1$ is not idempotent, since applying M_1 to $U_1 M_1$ yields the zero sequence and therefore also $U_1 M_1 (U_1 M_1 x) = 0 \neq U_1 M_1 x$.

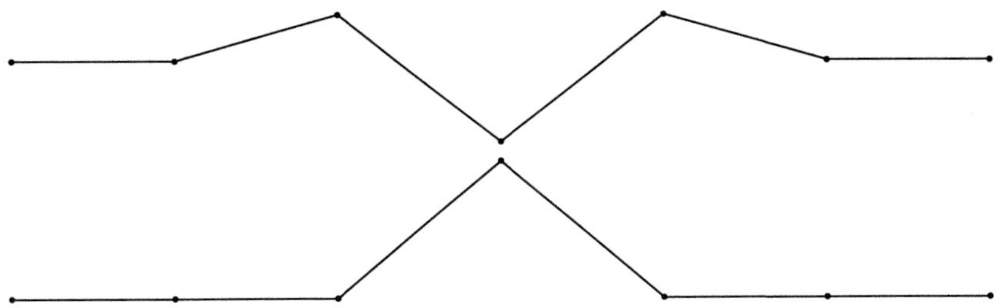

Figure 3.3. The operator $U_1 M_1$ applied to the sequence of three alternating inputs.

The operators UM^j and LM^j do however yield a root quickly (after two applications), as the following theorem demonstrates.

Theorem 3.8. For each $n \geq 0$, $(LnMn)^3 = (LnMn)^2$ and $(UnMn)^3 = (UnMn)^2$.

Proof.
$$\begin{aligned} (UnMn)(UnMn) &= Un(MnUn)Mn \\ &= Un(LnUn)Mn, && \text{since } MnUn = LnUn, \\ &= LnUnMn, && \text{since } UnLnUn = LnUn. \end{aligned}$$

Using this it follows that

$$(UnMn)(UnMnUnMn) = (UnMn)(LnUnMn).$$

Since $Un(MnLn)UnMn = Un(UnLn)UnMn = LnUnMn$, it follows that

$$(UnMn)^2 = (UnMn)^3 = LnUnMn.$$

This proves one half of the theorem, and a similar argument completes the proof of the theorem. □

Investigating the proof of the above theorem makes it clear that the operators $(UnMn)^2$ and $(LnMn)^2$ map any sequence onto an n-monotone sequence. The same is true for the operators $(UnMn^j)^2$ and $LnMn^j)^2$ as the following theorem demonstrates.

Theorem 3.9. For each $n \geq 0$ and $j \geq 0$,

$$(UnMn^j)^2 = (UnMn^j)^3 \quad \text{and} \quad (LnMn^j)^2 = (LnMn^j)^3.$$

Proof.
$$\begin{aligned}(UnMn^j)^2 &= Un(Mn^jUn)Mn^j = UnMn^{j-1}(MnUn)Mn^j \\ &= UnMn^{j-1}(LnUn)Mn^j.\end{aligned}$$

A standard induction argument proves the crucial equality

$$(UnMn^j)^2 = UnLnUnMn^j = LnUnMn^j.$$

For each $x \in X$, $LnUn(Mn^jx) = LnUn(y)$ is n-monotone, by Theorem 2.5. Since Mn preserves such sequences, $UnMn^j$ preserves $LnUny$ as well, proving one half of the theorem.

The other half is proved similarly, or using an argument with duals. □

Theorem 3.10. Let $m = \max\{n, k\}$. Then

$$UmLm \leq UnLnUkLk \leq LnUnLkUk \leq LmUm.$$

Proof. The operators are all syntone and $LU \geq UL$ so that

$$LnUn(LkUk) \geq LnUnUkLk \geq UnLnUkLk.$$

Suppose that $m = n$. Then

$$\begin{aligned}LnUnLkUk &\geq LnUnLnUk, && \text{because } Lk \geq Ln, \\ &\geq LnUnLn, && \text{because } Uk \geq I, \\ &= UnLn, && \text{by Theorem 2.9.}\end{aligned}$$

Similarly

$$\begin{aligned}LnUnLkUk &\leq LnUnIUk, && \text{since } Lk \leq I, \\ &\leq LnUnUn, && \text{since } Uk \leq Un, \\ &= LnUn, && \text{since } Un \text{ is idempotent.}\end{aligned}$$

A similar argument when $m = k$ completes the proof of the theorem. □

The above theorem refines the order relation on the set of selectors considerably, and yields a proof that several other classes of selectors and compositions map into sets of locally monotone sequences, and do so by mapping a sequence x into a sequence between $UmLmx$ and $LmUmx$. Examples are given by the following corollary.

Corollary. If $m = \max\{n, k\}$ then

$$UmLm \leq M_n^j M_k^i \leq LmUm \quad \text{for all} \quad i, j > 0, \quad \text{since}$$
$$UnLn \leq M_n^i \leq LnUn \quad \text{and} \quad UkLk \leq M_k^i \leq LkUk.$$

Apart from the popular smoothers Mn^∞ and Mn^*, there are now a whole class of smoothers composed of various $LnUn$ and Mn that map consistently onto the class $\mathcal{M}_n$ of n-monotone sequences. They are thus all effective, and since they are all idempotent, the criteria that can now be used to select from these are efficiency and stability. Consistency does however have other criteria, apart from idempotence and preservation of signals. One of these is co-idempotence.

There is however a philosophical problem to address. Do we really need to have an n-monotone sequence as output? This question is closely linked to the question of ambiguity suggested by the fact that Ln and Un do not commute and therefore generally yield different n-monotone sequences. The same holds for the alternative smoothers mapping onto $\mathcal{M}_n$. They all preserve each other's output, thus they agree with the respective concept of what signals are, but often yield different signals. We consider the smoothers addressed by Theorems 2.14 and 2.15.

Theorem 3.11. For each $N \geq 0$ the operator Q_n has the following properties.

(a) $Ln \leq Qn \leq Un$.
(b) $QnLn = UnLn$ and $QnUn = LnUn$.
(c) $(LnQn)^2 = (LnQn)^3$ and $(UnQn)^2 = (UnQn)^3$.

Proof.

(a) $(Un + Ln - I)x \leq Unx$ for each x, since $Lnx - x \leq 0$. Similarly $Qn \geq Ln$.
(b) $(Un + Ln - I)Ln = UnLn + LnLn - Ln = UnLn$, since Ln is idempotent. Similarly $QnUn = LnUn$.
(c) $LnQn)^2 = Ln(QnLn)Qn = LnUnLnQn = UnLnQn$, $(LnQn)^3 = LnQn(UnLnQn) = Ln(QnUn)LnQn = Ln(LnUn)LnQn = LnUnLnQn = UnLnQn$. □

The above properties have a remarkable similarity with those of the median smoother M_n. This is extremely odd, since Q_n is generally not a selector.

A selector S is an operator such that $(Sx)_i = x_j$, for some j, but the sequence $z = \langle z_i = 2\delta_{i,j-2} - \delta_{i,j-1} + \frac{1}{2}\delta_{i,j} - \delta_{i,j+1} + 2\delta_{i,j+2} \rangle$ is such that $(U_3z)_j = 2, (L_3z)_j = -1$ and $Q_3z = 1 - \frac{1}{2} \neq z_k, \forall k$.

Letting $x = z + \frac{1}{2}\langle \delta_{i,j} \rangle$ shows that $x > z$, but $(Q_3x)_j < (Q_3z)_j$, and therefore that Q_3 is not syntone.

The following theorem demonstrates, that there is also considerable similarity between the average of $LnUn$ and $UnLn$, M_n and Q_n.

Theorem 3.12. The operator $Gn = \frac{1}{2}[LnUn + UnLn]$ has the following properties:

(a) $Ln \leq Gn \leq Un$.
(b) $GnLn = UnLn$ and $GnUn = LnUn$.
(c) $(LnGn)^2 = (LnGn)^3$ and $(UnGn)^2 = (UnGn)^3$.

Proof.

(a) $Ln \leq UnLn \leq Gn \leq LnUn \leq Un$.
(b) $GnLn = \frac{1}{2}[UnLn + LnUn]Ln = \frac{1}{2}[Un(LnLn) + Ln(UnLn)] = \frac{1}{2}[UnLn + UnLn] = UnLn$.

A similar argument proves that $GnLn = LnUn$. Finally,

(c) $(LnGn)^2 = Ln(GnLn)Gn = Ln(UnLn)Gn = UnLnGn$,
$(LnGn)^3 = LnGn(UnLnGn) = Ln(GnUn)LnGn = Ln(LnUn)LnGn = UnLnGn$. □

Corollary. $(LnGn)^2$ maps X onto $\mathcal{M}_n$, since, for each $x \in X$, the sequences $UnLnGn(x) = UnLn(Gn(x)) \in \mathcal{M}_n$. The same is true for $(U_nGn)^2$.

Having a selection of smoothers available that map onto $\mathcal{M}_n$, is more than could be hoped for. It is thus possible to investigate and compare them as to their merit in secondary considerations. Effectiveness and some consistency is guaranteed, and the important criteria that remain are stability and efficiency. Stability seems to be a strong priority and can be addressed by demonstrating one unwanted instability in the popular smoothers mapping M^∞ and M^* onto $\mathcal{M}_n$. Both have a particularly annoying instability due to the fact that they are not sufficiently local. For each n, both M_n^∞ and M_n^* can undergo radical changes in sequences $M_n^\infty x$ and $M_n^* x$ at a particular index i, by having the sequence x changed at another index j, arbitrarily far away.

This is illustrated with two examples differing at only one value in the following figures.

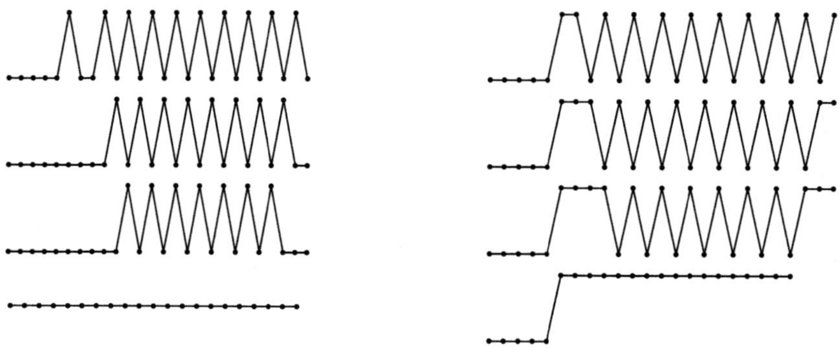

Figure 3.4. An illustration of the arbitrarily distant error influence with M^∞.

In the next chapter it will be seen that such an occurrence is rare, and not as serious as it may seem. As far as values of the smoothed sequence are concerned, it will be seen later that all the operators will be Lipschitz-continuous so that perturbations in x will not have disproportionate perturbations in the image under smoothing.

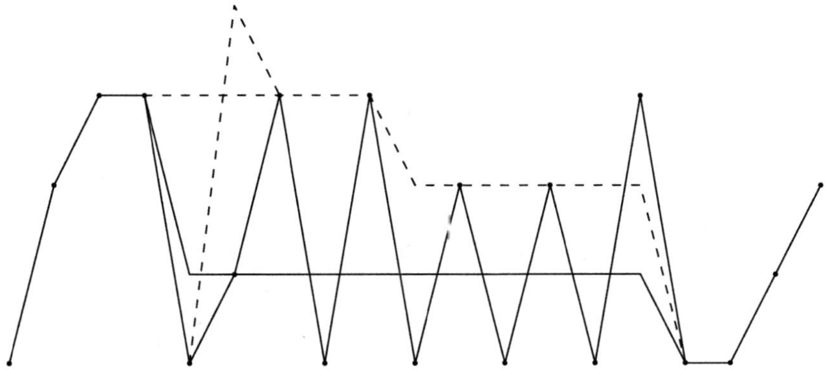

Figure 3.5. An illustration of the arbitrarily distant error influence with M^*.

Of academic interest is that repeated application of Gn to a sequence x normally quickly produces a unique unbiased n-monotone estimate of the signal in x. The convergence in practice is much faster than that of repeated application by Mn. The guaranteed rate of convergence is given by the following theorem.

Theorem 3.13. $||LnUnGnx - UnLnGnx||_\infty \leq \frac{1}{2}||LnUnx - UnLnx||_\infty$.

Proof. Let $w = LUx - ULx$, $c = \langle ||w||_\infty \rangle$ and $z = \frac{1}{2}(LUx + ULx) = ULx + \frac{1}{2}(LUx - ULx)$.

$LUz = LU(ULx + \frac{1}{2}w) \leq LU(ULx + \frac{1}{2}c)$, since $c \geq w$ and LU is syntone.

$ULz = UL(ULx + \frac{1}{2}w) \geq UL(ULx)$, since $w \geq 0$ and UL is syntone.

Therefore $LUz - ULz \leq LU(ULx) + \frac{1}{2}c - UL(ULx)$, and since $LUUL = UL$ is idempotent it follows that $0 \leq LUz - ULz \leq \frac{1}{2}c$.

But then $||LUz - ULz||_\infty \leq \frac{1}{2}||w||_\infty$, proving the theorem. □

Remark. The constant $\frac{1}{2}$ is the best possible as the following example shows.

Example. Let $e = \langle (-a)^i \rangle$.

$$Ue_i = \min\{\max\{(-a)^{i-1}, (-a)^i\}, \max\{(-a)^i, (-a)^{i+1}\}\}$$
$$= \begin{cases} a^i & \text{for } i \text{ even} \\ a^{i-1} & \text{for } i \text{ odd} \end{cases}.$$
$$Le_i = \begin{cases} -a^{i-1} & \text{for } i \text{ even} \\ -a^i & \text{for } i \text{ odd.} \end{cases}$$

Both are monotone sequences so that $ULe_i = Le_i$ and $LUe_i = Ue_i$.

$0 \leq LUe_i - ULe_i = a^i + a^{i-1} = a^i(1 + \frac{1}{a})$.

If $z_i = \frac{1}{2}(LUe_i + ULe_i) = (-1)^i a^i \frac{(1-\frac{1}{a})}{2}$, then

$$LUz_i - ULz_i = \frac{1}{2}\left(\frac{1-\frac{1}{a}}{1+\frac{1}{a}}\right) LUe_i - ULe_i.$$

Thus

$$||LUz - ULz||_p \leq \frac{1}{2}\left(\frac{a-1}{a+1}\right) ||LUe - ULe||_p,\ \forall p \geq 1.$$

For large a, $\frac{1}{2}\left(\frac{a-1}{a+1}\right)$ is arbitrarily close to $\frac{1}{2}$.

For the same example $(Mne)_i = (-a)^{i-1} = -\frac{1}{a}(-a)^i$. From this it is easy to show that

$$||M^{k+1}nx - M^k nx||_p = \frac{1}{|a|}\, ||M^k nx - M^{k-1}nx||,$$

demonstrating arbitrary slow convergence if $|a|$ is near 1. Computationally this is inconvenient, compared to G_n.

It will become apparent that the effectiveness, measured in terms of producing a signal, as defined, is not necessarily to be interpreted too strictly. Consistency can similarly be relaxed slightly, so as not to strictly require idempotence of the smoother. At this stage there are sufficiently many idempotent smoothers available, and also sufficiently many that preserve $\mathcal{M}_n$. The other half of a strict measure of consistency is the co-idempotence of the smoother. The *LULU*-operators will be shown to be co-idempotent and thus strictly consistent, but again this will be shown to permit relaxation.

The important measure of efficiency remains a consideration, but other, arguably relatively minor, considerations will still be addressed after the further consideration of consistency, using the concept of the *LULU*-interval.

A friend, Johannes Cronje, pointed out that in Quantum Mechanics the Heisenberg Uncertainty Principle comes into play precisely when the commutators of observables are nonzero. With this image in mind several problems in nonlinear smoothing, and later in the Multiresolution Analysis, can be seen in a clearer perspective. The Heisenberg Uncertainty Principle is central in the mapping between a function and its Fourier Transform, and a function and its Wavelet Transforms. A different, but similar, uncertainty may be fundamental between a sequence and its Pulse Transforms, as defined later. This idea turns out to be a guide towards useful and beautiful results.

A further obvious idea is to use secondary attributes to select amongst the "equivalent" smoothers inside the *LULU*-interval. It is in the selection of these secondary attributes that more insight is required. Some surprisingly useful ones may present themselves.

4. *LULU*-Intervals, Noise and Co-idempotence

Vom Unglück erst
Zieh ab die Schuld
Was übrig ist,
trag in Geduld.
Theodor Storm

Choosing the n-monotone sequences as signals is not only convenient, being precisely the range of the *LULU*-operators $LnUn$ and $UnLn$, but also appropriate. The n-monotone sequences are precisely those that have no upward pulse of width not more than n, if this is measured relative to the neighboring values. What is a philosophical puzzle, is why there should be such natural idempotent mappings onto $\mathcal{M}_n$, that came in unequal pairs. Noting that the inequality $LnUn \geq UnLn$ is strict, and should be written as $LnUn > UnLn$, it is clear that there generally is a pair of equally valid n-monotone "approximations" of a sequence $x \in X$. This ambiguity turns out to be natural and significant rather than an oddity of logic, when interpreted heuristically. There is a natural ambiguity inherent in the definition of impulsive noise, which can be interpreted as follows.

Consider the sequences $y = \langle x_i = \delta_{i,j} + \delta_{i,j-1} + \delta_{i,j+1} \rangle$ and $z = 0$ in $\mathcal{M}_1$. How is the sequence $x = \langle x_i = \delta_{i,j-1} + \delta_{i,j+1} \rangle$ to be interpreted. Is it y with a single downward impulse at j or z with a pair of almost adjacent impulses upward? Both are equally valid simple interpretations. This can be considered a fundamental ambiguity.

Experimentally, the ambiguity is very illuminating. If a smooth function is sampled, reasonably distributed (near Gaussian) noise added, and impulses, to yield a sequence x, then a norm of the difference between $LnUnx$ and $UnLnx$ seems to yield a good indication of the "amount" of well distributed noise. When there is no reasonable ambiguity, such as when there is an isolated n-blockpulse on a constant sequence c, the interval $[UnLnx, LnUnx]$ is precisely the interval $[c, c]$. It is to be noted that, for $n > 1$, refined ambiguity interpretations could be argued for generally narrower "intervals of ambiguity", as for example $U_2L_2U_1L_1x$ and $L_2U_2L_1U_1x$ which, by Theorem 3.10, lie in the interval between U_2L_2x and L_2U_2x. It is convenient to consider the difference as a robust indicator of the uncertainty in interpretation. The maximum ambiguity is attained when a periodic section

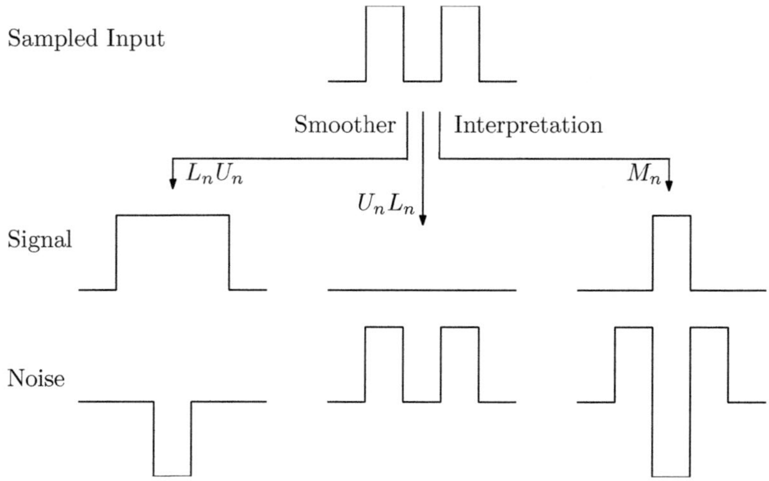

Figure 4.1. The fundamental ambiguity in the concept of impulsive noise.

is sampled at precisely the Nyquist-frequency. As an example it is illustrative to sample a function of the type $f(t) = \sin(\alpha t + \beta)$, which has a linearly increasing frequency α. For low frequencies the difference between LUx and ULx is very small, where $x = \langle x_i = f(h_i)\rangle$. As i increases, so does the frequency. Progressively the interval between LUx and ULx grows until it is an envelope at the Nyquist frequency, after which we have the familiar aliasing effect. This is well known to signal analysts, anybody familiar with digital filters and with the Shannon–Whitney sampling theorem.

A constant sequence has an impulse as a Fourier Transform and conversely, an impulse has a constant Fourier Transform. These are extremes of the general uncertainty in the essential supports of a function and its Fourier Transform. If a function is narrow in time support, it is wide in frequency content. This can be more precisely formulated, and is known as Heisenberg's uncertainty principle (8), and is central in the analysis of Windowed Transforms, and in the wide range of Wavelet Transforms receiving considerable attention currently.

Aliasing is not restricted to trigonometric functions either, but to any sampling for linear approximation by functions from a subspace of functions, such as polynomials of degree m. Thus even when, as often in the case of linear signal processing, high frequencies are considered noise, and low frequencies signal, the *LULU*-operators indicate the danger of aliasing by the fundamental separation at multiples of the Nyquist frequency. There clearly is significant room for further interpretation, but to obtain more insight it is necessary to formalize the ideas further.

Definition. The closed interval $[UnLnx, LnUnx]$ is called the n-signal interval of $x \in X$ and denoted by $[\underline{x}, \overline{x}]_n$, or $[\underline{x}, \overline{x}]$ when n is implied.

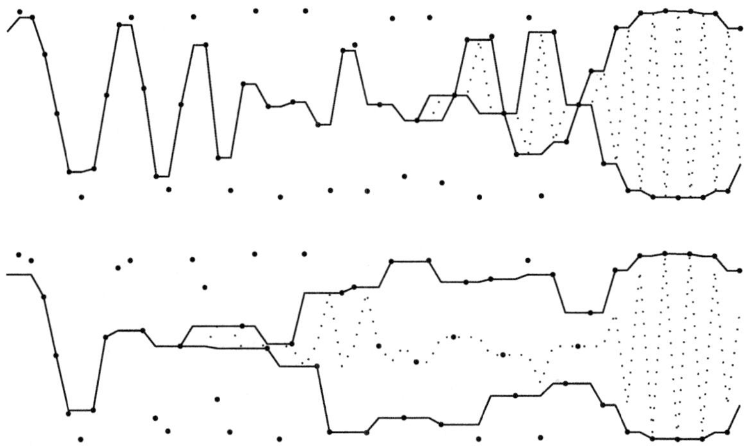

Figure 4.2. A sampling x of a sinusoidal with increasing frequency and the *LULU*-images. The median interpretation is contained in between.

Remarks.

(a) $[\underline{x}, \overline{x}]_0$ can be identified with x itself.

(b) $[\underline{x}, \overline{x}]_n$ is identified with the sequence x if $x \in \mathcal{M}_n$.

The definition of the *LULU*-interval , or n-signal interval (n-interval), can be expanded to map sequence intervals $[x, y]$, with $y \leq x$, onto $[UnLnx, LnUny]$. Also it is useful to define operator intervals like $[UL, LU]$.

Definition. Let $x \leq y$ be sequences from X. Then the *n-LULU* interval of $[x, y]$ is $[UnLn(x), LnUn(y)]$.

Definition. The interval $[A, B]$ can be defined to be the set of all operators C such that $A \leq C \leq B$.

Definition. $\langle \underline{x}, \overline{x} \rangle_n = [\underline{x}, \overline{x}]_n \cap \mathcal{M}_n$ is the set of all n-monotone sequences in $[\underline{x}, \overline{x}]_n$.

All the above concepts permit a consistent framework for the comparison of different smoothers, amongst them the popular "repeated till convergence" operator $R_n = M_n^{\infty}$. (In practice the sequences that are to be smoothed are finite and can be appended with zeroes to be bi-infinite, yet permit the existence of R_n.) Another popular smoother, the recursive median M_n^*, is also comparable. To make these concepts precise, the following development is expedient.

Since the ultimate objective is not to produce an n-monotone sequence as such, but to obtain a sequence that is sufficiently free of impulsive noise of a prescribed type, in the precise sense of being within the primary "interval of ambiguity" as to interpretation of impulsive noise, it is not important whether R_n or M_n^* produce n-monotone sequences or not. There are other good smoothers as well that do not. It would therefore be sufficient if an operator maps a general sequence x <u>into</u> the interval $[\underline{x}, \overline{x}]_n$. This idea prompts the following.

Definition. A sequence interval $[a, b]$ is included in $[c, d]$ if $c \leq a \leq b \leq d$ and is denoted by $[a, b] \subset [c, d]$.

Theorem 4.1. Interval inclusion is an order relation.

Proof.

(a) $[a, b] \subset [a, b]$.
(b) $[a, b] \subset [c, d]$ and $[c, d] \subset [a, b] \Rightarrow [a, b] = [c, d]$.
(c) $[a, b] \subset [c, d]$ and $[c, d] \subset [e, f] \Rightarrow [a, b] \subset [e, f]$. □

Definition. Two sequences x and y are n-*LULU* equivalent if $[\underline{x}, \overline{x}]_n = [\underline{y}, \overline{y}]_n$.

Theorem 4.2. *LULU*-equivalence is an equivalence relation.

Proof. The relation is easily proven to be reflexive, symmetric and transitive. □

The *LULU*-equivalence clearly partitions X into different "ancestor" classes; all those sequences mapping onto the same *LULU*-interval. Noise added to $x \in X$ can be considered removable if it keeps the sequence in the same ancestor class. This is a less restrictive concept of removable noise than that used by Callager and Wise, who consider removable noise as that which changes a sequence into one that has the same "root". Yet even this concept is too restrictive, since it would be acceptable if the addition of noise only keeps the new *LULU*-interval included in the old one.

In a similar spirit an operator A could be called n-*LULU*-similar (n-similar) if it merely is such that $A \subset [UnLn, LnUn]$, and a sequence z called n-*LULU*-similar to x if $z \in [\underline{x}, \overline{x}]$. (These relations are not equivalence relations.)

Several important smoothers are n-*LULU*-similar. Included are the operators

$$R_n = M_n^k, \quad M_n^*, \quad \frac{1}{2}[UnLn + LnUn], \quad \text{and so is} \quad Q_n \quad \text{if} \quad n = 1 \text{ or } 2.$$

The operator $M_2M_1^2$, popular with Tukey, is specifically 2-*LULU*-similar.

Theorem 4.3. Let A_n be an n-*LULU*-similar operator. Then A_n annihilates n-impulses and preserves n-monotone sequences.

Proof. Let x be an n-impulse. Then $0 = UnLn(x) \leq A(x) \leq LnLn(x) = 0$. If x is n-monotone, then similarly $x = UnLn(x) \leq A(x) \leq LnUn(x) = x$. □

To prove the similarity for some of the important smoothers, the following theorems are required.

Theorem 4.4. For each integer $k \geq 1$, M_n^k and G_n^k are n-*LULU*-similar.

Proof. The smoothers Un, Ln and Mn are all syntone, so that all compositions are syntone. Noting that $UnLn \leq Mn \leq LnUn$, it follows that

$$(UnLn)^2 \leq UnLnMn \leq MnMn \quad \text{and} \quad (LnUn)^2 \geq LnUnMn \geq MnMn.$$

Since $LnUn$ and $UnLn$ are idempotent, $UnLn \leq Mn^2 \leq LnUn$.

The proof is completed by induction. The proof for G_n^k is similar. □

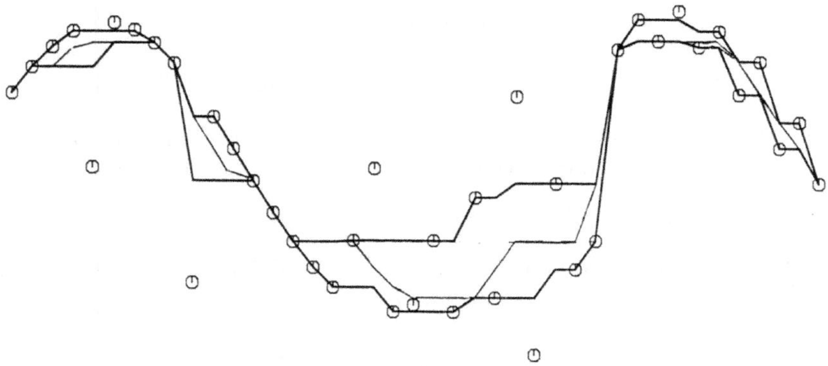

Figure 4.3. The smoother R_n^∞ mapping x into the $LULU$-interval

Corollary. If $R_n^\infty x$ exists it is also in $[UnLnx, LnUnx]$.

To prove that the recursive medians are $LULU$-similar requires the following theorem and its required definitions. (The result is remarkable, to anyone familiar with recursive digital filters.) To show this it is required to define the recursive variants, for which the starting procedure is important. It is therefore assumed that the sequences used in practice are finite, so that somewhere in the past (for sufficiently small index) the sequences have zero elements only.

Definition. $U^*n(x)_i = Un(y)_i$, where $y_j = \begin{cases} x_j, & \text{for } j \geq i, \\ U^*n(x)_j, & \text{for } j < i. \end{cases}$

$$L^*n(x)_i = Ln(y)_i, \text{ where } y_j = \begin{cases} x_j, & \text{for } j \geq i, \\ L^*n(x)_j, & \text{for } j < i. \end{cases}$$

$$M^*n(x)_i = Mn(y)_i, \text{ where } y_j = \begin{cases} x_j, & \text{for } j \geq i, \\ M^*n(x)_j, & \text{for } j < i. \end{cases}$$

Theorem 4.5. $U^*n = Un$ and $L^*n = Ln$.

Proof. There is an integer s such that $U^*n(x)_i = Un(x)_i = x_i = 0$ for all $i \leq s$. Suppose that $U^*n(x)_i = Un(x)_i$ for all $i \leq k$. $U^*n(x)_{k+1} = Un(y)_{k+1}$, where

$$y_i = \begin{cases} x_j, & \text{for } j \geq k+1, \\ U^*n(x)_j, & \text{for } j < k+1. \end{cases}$$

It is clear that $y \geq x$ yields $U^*n(x)_{k+1} \equiv Un(y)_{k+1} \geq Un(x)_{k+1}$. But also $y \leq Unx$ yields $U^*n(x)_{k+1} = Un(y)_{k+1} \leq Un(Unx)_{k+1} = Un(x)_{k+1}$. □

The proof is completed by the standard induction argument.

Lemma. $Ln \leq M^*n \leq Un$.

Proof. There is an index s such that $Ln(x)_i \leq M^*n(x)_i \leq Un(x)_i$, for all $i \leq s$. Suppose that $M^*n(x)_k \leq Un(x)_k$. Then

$$M^*n(x)_{k+1} = Mn(y)_{k+1}, \quad \text{with} \quad y_j = \begin{cases} x_j, & j \geq i, \\ M^*n(x)_j, & j < i. \end{cases}$$

But, by the assumption, $y \leq Un(x)$, so that

$$M^*n(x)_{k+1} \leq Mn(Unx)_{k+1} \leq Un(Unx)_{k+1} = Un(x)_{k+1}.$$

A standard induction argument proves that $M^*n \leq Un$, and a similar argument to the above proves the rest of the lemma. □

Theorem 4.6. $M^*n \subset [UnLn, LnUn]$.

Proof. There is an s such that $0 = LnUn(x)_i = M^*n(x)_i = UnLn(x)_i$, for all $i \leq s$. Suppose that $M^*n(x)_k \leq LnUn(x)_k$.

$$M^*n(x)_{k+1} = Mn(y)_{k+1}, \quad \text{with} \quad y_i = \begin{cases} x_j, & \text{for } j \geq i, \\ M^*n(x)_j, & \text{for } j < i. \end{cases}$$

$Mn(y)_{k+1} \leq LnUn(y)_{k+1} \leq LnUn(Unx)_{k+1}$, since $Ux_n \geq y$.

This proves that $M^*n(x)_{k+1} \leq LnUn(x)_{k+1}$, and a standard induction argument proves that $M^*n(x) \leq LnUn(x)$.

A similar argument proves that $M^*n(x) \geq UnLn(x)$; since the two inequalities are true for all sequences concerned, the theorem is proved. □

The fact that the recursive median M^*n is in the interval $[UnLn, LnUn]$ is astonishing, though not much more so than the innocent looking fact that M^kn is also. This becomes clear when considering the supports of M^kn and M^*n. Note that the support of $LnUn$ and $UnLn$ is an index interval of $[i-2n, i+2n]$, and that of Mn is $[i-n, i+n]$. M^2n has a support interval $[i-2n, i+2n]$ and this increases with each additional power. Similarly, the support of the recursive median smoother is arbitrarily large, as was demonstrated by a previous argument for the instability of the recursive median. How can a local smoother bound a non-local one, or even just one with an arbitrarily larger support? It takes some careful thought to digest this seemingly paradoxical fact, which could not be possible in linear (digital) filters.

Two further examples of useful smoothers that are n-*LULU* similar are the following.

Definition. Let Bn be the operator defined by

$$Bn(x)_i = \begin{cases} x_i, & \text{if } x_i \in [UnLn(x)_i, LnUn(x)_i], \\ UnLnx_i, & \text{if } x_i < UnLn(x)_i, \\ LnUnx_i, & \text{if } x_i > LnUn(x)_i. \end{cases}$$

Theorem 4.7. $Bn \in [UnLn, LnUn]$.

Proof. For each $x \in X$, $UnLnx_i \leq Bn(x)_i \leq LnUn(x)_i$ for each index i. This means that $UnLn(x) \leq Bn(x) \leq LnUn(x)$, proving the theorem. □

An alternative to Bn is often useful, which is also n-*LULU*-similar, but in the real example below is somewhat better than Bn.

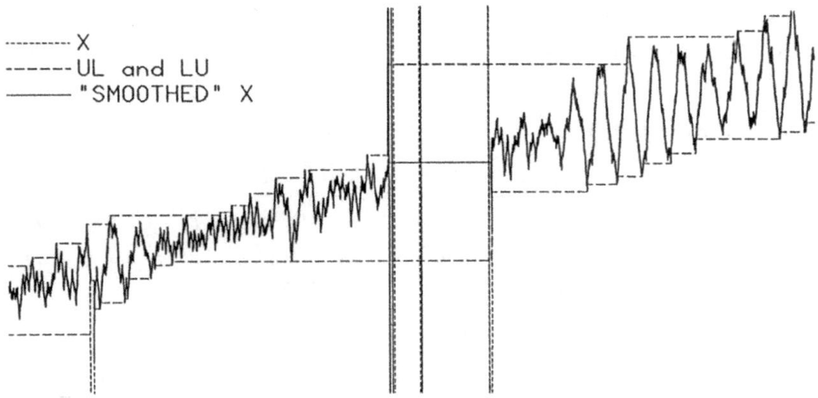

Figure 4.4. Removal of impulsive noise by the operator An : 6500 of 430 000 points from the trajectory of Figure 1 in the introduction. (n is large).

Definition. $An = \begin{cases} x_i, & \text{if } x_i \in [UnLnx_i, LnUnx_i], \\ \frac{1}{2}[ULnx_i + LnUnx_i], & \text{otherwise.} \end{cases}$

For the sake of consistency it was argued that an operator should ideally be idempotent and co-idempotent. The demand of the idempotence has been shown to be needlessly strict, since it could be argued that it is sufficient to be within an interval of ambiguity provided by consistent operators, in this case the appropriate *LULU*-operators. A similar relaxation could be sufficient for co-idempotence, but this needs to be investigated.

The smoothers Un and Ln can be shown to be co-idempotent by the following theorems.

Theorem 4.8. $L_n(I - L_n) \geq 0$ and $U_n(I - U_n) \leq 0$.

Proof. $I - L_n \geq 0 \geq I - U_n$ since $L_n \leq I \leq U_n$. Since L_n and U_n are syntone it follows that

$$L_n(I - L_n) \geq L_n 0 = 0 \quad \text{and} \quad 0 = U_n 0 \geq U_n(I - U_n).$$

The above theorem seems to be as far as one can get, using only the properties listed above, to proving the co-idempotence of L_n and U_n. To complete the proof of the co-idempotence seems to require an operator specific proof of the following theorem.

Theorem 4.9. $L_n(I - L_n) \leq 0$ and $U_n(I - U_n) \geq 0$.

Proof. Suppose $L_n(I - L_n)x_i > 0$. Using the notation $L = L_n$ and $u_j = x_j - Lx_j$, this means that $\max\{\min\{u_{i-n}, \ldots, u_i\}, \ldots, \min\{u_i, \ldots, u_{i+n}\}\} > 0$. Thus $\min\{u_j \ldots, u_{j+n}\} > 0$ for at least one $j \in [i-n, i]$, or; $x_j - Lx_j > 0, x_{j+1} - Lx_{j+1} > 0, \ldots, x_{j+n} - Lx_{j+n} > 0$ which means that $x_k > Lx_k$ for all $k \in [j, j+n]$. But

$$Lx_k = \max\{\min\{x_{k-n}, \ldots, x_k\}, \ldots, \min\{x_k, \ldots, x_{k+n}\}\} \geq \min\{x_j, \ldots, x_{j+n}\}.$$

This means there is a value x_s in $\{x_j, \ldots, x_{j+n}\}$ which is not larger than any of the values Lx_k, and thus also Lx_s. This is a contradiction. A similar proof yields $U_n(I - U_n) \geq 0$. □

The following partial result is a simple consequence of the above results.

Theorem 4.10. $U_nL_n(I - U_nL_n) \leq 0$ and $L_nU_n(I - L_nU_n) \geq 0$.

Proof. $U_nL_n(I - L_n) = 0$, since $L_n(I - L_n) = 0$. Since $I - U_nL_n \leq I - L_n$ it follows, because U_nL_n is syntone, that $U_nL_n(I - U_nL_n) \leq U_nL_n(I - L_n) = 0$.

A similar proof holds for the other part of the theorem. □

Theorem 4.11. $U_nL_n(I - U_nL_n) = 0$ and $L_nU_n(I - L_nU_n) = 0$.

Proof. With $U = U_n$ and $L = L_n$, assume that $LU(I - LU)x_i > 0$ for some x and i. With the notation $w_j = U(I - LU)x_j$, this means that

$$\max\{\min\{w_{i-n}, \ldots, w_i\}, \ldots, \min\{w_i, \ldots, w_{i+n}\}\} > 0.$$

Assume that $\min\{w_{k-n}, \ldots, w_k\} > 0$, with $k \in [i, i+n]$. This implies that $U(I - LU)x_\ell > 0$ for each $\ell \in [k-n, k]$.

With the notation $v_j = (I - LU)x_j$, this means that

$$\min\{\max\{v_{\ell-n}, \ldots, v_\ell\}, \ldots, \max\{v_\ell, \ldots, v_{\ell+n}\}\} > 0.$$

This means that $x_j > LUx_j$ for one value j in each of the sets $\{k-2n, \ldots, k-n\}, \ldots, \{k, \ldots, k+n\}$.

Let $\varepsilon = x_m = \min\{x_j; x_j > LUx_j, k - 2n \leq j \leq k+n\}$. For each $i \in [k-2n, k+n]$ it follows that

$$\begin{aligned} Ux_i &= \min\{\max\{x_{i-n}, \ldots, x_i\}, \ldots, \max\{x_j, \ldots, x_{i+n}\}\} \\ &\geq \min\{\varepsilon, \varepsilon, \ldots, \varepsilon\} = \varepsilon. \end{aligned}$$

But then $LUx_i = \max\{\min\{Ux_{i-n}, \ldots, Ux_i\}, \ldots, \min\{Ux_i, \ldots, Ux_{i+n}\}\} \geq \varepsilon$, and since this is also true for $i = m$ it follows that $LUx_m \geq \varepsilon > LUx_m$.

This contradiction, together with the above theorem, yields $LU(I - LU) = 0$.

A similar proof holds for $UL(I - UL) = 0$. □

Corollary.

(a) $L_n(I - U_nL_n) \leq 0$ and $U_n(I - L_nU_n) \geq 0$ since

$$\begin{aligned} &L_n(I - U_nL_n) \leq U_nL_n(I - U_nL_n) = 0 \\ \text{and } &U_n(I - L_nU_n) \geq L_nU_n(I - L_nU_n) = 0. \end{aligned}$$

(b) $L_nU_n(U_nL_n - I) = 0$ and $U_mL_n(L_nU_n - I) = 0$ since

$$\begin{aligned} 0 &= L_nU_n(I - L_nU_n)(-x) = L_nU_n(I(-x) - L_nU_n(-x)) \\ &= L_nU_n(-Ix - L_n(-L_nx)) = L_nU_n(-Ix + U_nL_nx). \end{aligned}$$

A similar proof holds for $U_nL_n(L_nU_n - I) = 0$.

All of the above inequalities and equalities can be verified by testing on randomly generated sequences. The really worthwhile results are that $L_nU_n(I - L_nU_n) = 0$ and $U_nL_n(I - U_nL_n) = 0$, since this implies L_nU_n and U_nL_n are co-idempotent.

Apart from the confirmation of consistency argued previously, one of the consequences is that L_nU_n and U_nL_n have only eigenvalues 0 and 1, since $L_nU_n(\lambda x) = \lambda L_nU_n(x)$ and $U_nL_n(\lambda x) = \lambda L_nU_n(x)$ for $\lambda > 0$. It is worthwhile proving this with the following general result. Recalling that an operator P is scale independent if $P(\lambda x) = \lambda Px$ for $\lambda \geq 0$ (third smoother axiom), we get;

Theorem 4.12. If P is a scale independent separator, then it has only the eigenvalues 1 and 0.

Proof. Let $Pe = \lambda e$.

(a) If $\lambda \geq 0$, then $P^2e = P(\lambda e) = \lambda Pe = \lambda^2 e = Pe = \lambda e$. But $(\lambda^2 - \lambda)e = 0$ implies $\lambda^2 - \lambda = 0$ or $\lambda = 1, 0$.

(b) If $\lambda < 0$, then $(I-P)e = e-\lambda e = (1-\lambda)e$. Since $P(I-P)e = 0 = P(1-\lambda)e = (1-\lambda)Pe = (1-\lambda)\lambda e$, it follows that $\lambda(1-\lambda) = 0$, which is impossible for $\lambda < 0$.

Corollary. The range of a scale independent separator P is the set of eigensequences w.r.t. 1, unless the range contains only the zero sequence. Similarly, the range of $I - P$ is the set of eigensequences of 0.

Since each sequence x can be decomposed into $x = Px + (I - P)x$ it is clear that, unless $P = I$ or $P = 0$, x can be decomposed into Px, the eigensequence w.r.t. the eigenvalue 1 ("signal") and $(I - P)x$, the eigensequence w.r.t. the eigenvalue 0 ("noise"). It is important to note that, if $x = y + z$, with y in the range of P, and therefore a "signal", it does not follow that $y = Pz$ and $z = (I - P)x$. This can be seen from the examples L_nU_n and U_nL_n, which are not identical, though their ranges are precisely the set of n-monotone sequences, and are therefore identical. What is significant is that;

$$\text{Range}(I - L_nU_n) \neq \text{Range}(I - U_nL_n).$$

Even when x is such that $L_nU_nx = U_nL_nx$, the decomposition of x into an n-monotone sequence and an eigensequence w.r.t. the eigenvalue 0 is not unique in general, as is demonstrated by the examples $x = \langle(i < 0)\rangle$, $w = \langle(i < 1)\rangle$ and $v = \langle(i < 2)\rangle$, which are all monotone and $w - x$ and $v - w$ are such that $L_1U_1(w - x) = U_1L_1(w - x) = L_1U_1(v - w) = U_1L_1(v - w) = 0$.

By choosing counterexamples the operators M^kn, Qn and Gn can be shown to be not co-idempotent. The other operators $M^*n, M^\infty n, UnLnQn, LnUnQ$ etc have not been investigated. There does not seem to be a need for this, in light of the fact that there is an interval of uncertainty associated with the noise, as determined by $LnUn$ and $UnLn$. They confirm each other's interpretation of signal but not noise. Heuristically, it seems sufficient to have the residual of an extracted signal within this interval of ambiguity, as interpreted by the consistent pair $I - LnUn$

and $I - UnLn$. It is clear that $I - LnUn \leq I - UnLn$, so that it is reasonable to consider the following definition.

Definition. The n-noise interval of a sequence x is $[(I - LnUn)x, (I - UnLn)x]$.

Theorem 4.13. Let A_n be an operator such that $A_n \in [UnLn, LnUn]$. Then $I - A_n \in [(I - LnUn)x, (I - UnLn)x]$.

Proof. If $UnLnx \leq Anx \leq LnUnx$, then $-UnLnx \geq -Anx \geq -LnUnx$ and $(I - UnLn)x \geq (I - An)x \geq (I - LnUn)x$. Since this is true for each $x \in X$, the theorem is proved. □

Since $I - An - (I - An)^2 = An(I - An)$, the ambiguity as to the amount of signal still left in the residual $(I - An)x$ can be considered to be in $[UnLn(I - LnUn)x, LnUn(I - UnLn)]$, which is the union of two *LULU*-intervals.

Since it has been proved that $U_2L_2 \leq U_2L_2U_1L_1 \leq L_2U_2L_1U_1 \leq L_2U_2$, it follows that $[U_2L_2U_1L_1,\ L_2U_2L_1U_2] \subset [U_2L_2, L_2U_2]$, so that a narrower interval of ambiguity exists inside the 2-interval.

Similarly it may be of use to define operators C_n and F_n recursively as follows:

Definition. $C_1 = L_1U_1$ and $C_{n+1} = L_{n+1}U_{n+1}C_n$, $F_1 = U_1L_1$ and $F_{n+1} = U_{n+1}L_{n+1}F_n$.

Theorem 4.14. For each n; $UnLn \leq Fn \leq Cn \leq LnUn$.

Proof. The theorem is true for $n = 1$. Assume it is true for $n = k$, or $U_kL_k \leq F_k \leq C_k \leq L_kU_k$. But

$$\begin{aligned}
U_{k+1}L_{k+1} &\leq U_{k+1}L_{k+1}U_kL_k, && \text{by Theorem 3.10,}\\
&\leq U_{k+1}L_{k+1}F_k, && \text{by assumption,}\\
&\leq U_{k+1}L_{k+1}C_k \\
&\leq L_{k+1}U_{k+1}C_k \\
&\leq L_{k+1}U_{k+1}L_kU_k, && \text{by assumption,}\\
&\leq L_{k+1}U_{k+1}, && \text{by Theorem 3.10.}
\end{aligned}$$

The standard induction argument completes the proof. □

Looking at an example will demonstrate the idea that the operators C_n and F_n may be more realistic smoothers than L_n and U_n for the removal of pulses of length not exceeding n. Let

$$x = \langle x_i = \delta_{i,j} + \delta_{i,j+5} \rangle.$$

Clearly x is a zero sequence with two single impulses at j and at $j + 5$. U_6L_6 and L_6U_6 give outputs that are both 0 everywhere, U_5L_5 and L_5U_5 give different outputs.

$U_5L_5x = 0$, but L_5U_5 is a 5-blockpulse stretching from j to $j + 5$. The interpretation can be made that, if the noise consists of randomly located impulses,

L_5U_5x implies that the signal is a pulse of width 5 with 4 negative blockpulses at $j+1, j+2, j+3$ and $j+4$. This is a less probable occurrence than that assumed by U_5L_5; that the signal was zero with two single pulses located coincidentally at j and $j+5$. Clearly the expected distribution of impulses will determine the design of an appropriate smoother from smoothers that are compositions of U_jL_j and L_jU_j with $j \leq n$ rather than $UnLn$ and $LnUn$, to yield more likely interpretations if pulses of briefness not exceeding n are considered to have to be removed.

A choice that yields a small interval of ambiguity (interpretation) seems $[F_n, C_n]$, although there may be subtle arguments to include other, previously discussed smoothers in the *LULU*-interval.

A further reason for choosing the smoothers F_n and C_n will appear later, in connection with the problem of decomposing a given sequence into pulses of different resolution levels – Multiresolution Analysis (MRA) with (block-) pulses. Interestingly, the apparent increase in computational effort to compute F_n and Cn as opposed to simpler smoothers like $UnLn$ and $LnUn$ turns out to be misleading.

5. Smoothing and Approximation with Signals

Ich lebe mein Leben in wachsenden Ringen
die sich über die Dinge ziehn.
Ich werde den letzten vielleicht nicht vollbringen
aber versuchen will ich ihn.

Rainer Maria Rilke

The set $\mathcal{M}_n$ of n-monotone sequences is clearly of crucial importance in the subject of nonlinear smoothing. Many smoothers map onto or into $\mathcal{M}_n$. Is there a possibility of selecting any that are optimal in some sense?

Considering the operator B that maps x onto a "nearest" element in $\mathcal{M}_n$, in the sense of the metric chosen, the questions that come to mind are: For which metric does B exist, and if it maps onto an interval of sequences, can a suitable choice be made to define an unambiguous n-monotone image? How does the $LULU$-interval compare with the best?

The questions come from experience in Approximation Theory. Given a function f, we seek a best approximation Bf in a subspace S of approximating functions. For some norms the function Bf, though uniquely defined, is computationally complex. The idea then is to use a simpler operator P that is computationally efficient, like an interpolant, and use the Lebesgue inequality to show that this sub-optimal approximation Pf is not "far" from the best approximation Bf. Thus Pf suffices for practical purposes, or can possibly be used as a good start for an iterative procedure converging to Bf.

Restrepo and Bovic did a logical thing by considering the problem of best approximation in all the usual p-norms and seminorms as well $(0 < p < 1)$. This idea had been relatively unexplored, but seems sound. The p-norms are mappings from $X \times X$ to $\mathbb{R}$ are positive definite and symmetric and thus form a semi-metric. They are positive homogenous $(d_p(\gamma x, \gamma y) = |\gamma| d_p(x,y))$ and translation invariant. For $p \geq 1$ a metric is obtained, but for $0 < p < 1$ the triangle inequality does not hold.

They make a case for the semi-metric, but it is sufficient here to consider only the cases $p \geq 1$. They prove the existence of and suggest computational procedures for obtaining such "best" locally monotone sequences. The computational complexity is however uninviting and non-local.

As in the case of function approximation, it is thus natural to seek simple, possibly local, suboptimal methods that yield sequences in $\mathcal{M}_n$ that are near the optimal p-projection image, or best approximation. In linear approximation, the Lebesgue inequality provides such a comparison method, provided the operator norm is small. But in the problem at hand the operators that are candidates cannot be linear for previously argued reasons. It is however worthwhile to pursue the idea for possible generalization. It may even be speculated that the best approximation to a sequence x lies in the $LULU$-interval! This turns out to be not true in general, but still possibly true for a large class of sequences. When this is not so, this also may actually reflect more on the best approximation than on the $LULU$-interval!

The Lebesgue inequality requires that the operator is linear, idempotent and of bounded norm. Is the idea of the Lebesgue inequality transferable to the problem at hand; where the approximating set $\mathcal{M}_n$ is not a subspace and the operators not linear, though often idempotent? It is useful to review the Lebesgue inequality and a proof for the purpose of possible analogy.

Theorem. Let S be a subspace of a normed space X and P a linear mapping into S that is the identity on S and has bounded induced norm $\mu = ||P||$. Then, for each $x \in X$ and $s \in S$. $||Px - x|| \leq (||P|| + 1)||x - s||$.

Proof.

$$\begin{aligned} ||Px - x|| &\leq ||Px - s|| + ||s - x|| && \text{(Triangle inequality)} \\ &\leq ||Px - Ps|| + ||s - x|| && \text{(P is identity on S)} \\ &\leq ||P(x - s)|| + ||s - x|| && \text{(P is linear)} \\ &\leq ||P||\,||x - s|| + ||s - x|| && \text{(Induced norm)} \\ &\leq (1 + \mu)||s - x||. \end{aligned}$$

□

Replacing the subspace S by $\mathcal{M}_n$, a subset in X, it is clear that the triangle inequality is still valid, and the linearity demand on P can be circumvented by directly finding a Lipschitz-constant μ for P. Providing that P maps X onto $\mathcal{M}_n$ and has a Lipschitz-constant μ, the comparison between Px and any s in $\mathcal{M}_n$ is still valid, and thus Px can be comparable with the "best" sequence in $\mathcal{M}_n$. Considering the set of Lipschitz constants for an operator on X and choosing its minimum yields a minimum Lipschitz constant, which is a mapping μ onto $\mathbb{R}$, the real numbers, defined by;

Definition. (minlip constant)

$$\mu(P) = \sup \left\{ \frac{||Px - Py||}{||x - y||},\ x, y \in X,\ x \neq y \right\}.$$

The minlip mapping μ has the properties;

i) Clearly $\mu(C) = 0$, where C is a constant mapping $C(x) = c$. Conversely; let $\mu(P) = 0$. Then, $||Px - Py|| = 0$, $\forall x, y \in X$. Thus, $Px = Py = c \in X$.

ii)
$$\mu(\alpha P) = \sup\left\{\frac{||\alpha Px - \alpha Py||}{||x-y||}, \quad x, y \in X,\ x \neq y\right\}$$
$$= \sup\left\{|\alpha|\frac{||Px - Py||}{||x-y||}\right\}$$
$$= |\alpha| \sup\left\{\frac{||Px - Py||}{||x-y||}\right\} = |\alpha|\mu(P).$$

iii)
$$\mu(P+A) = \sup\left\{\frac{||(P+A)x - (P+A)y||}{||x-y||}\right\}$$
$$\leq \sup\left\{\frac{||Px - Py||}{||x-y||} + \frac{||Ax - Ay||}{||x-y||}\right\}.$$

But then $\mu(P+A) \leq \sup\{\mu(P) + \mu(A)\}$, since $\mu(P)$ and $\mu(A)$ are upper bounds and $\mu(P+A) \leq \mu(P) + \mu(A)$.

The two idempotent smoothers LU, UL are candidates for comparison with the best approximation in $\mathcal{M}_n$. The following lemmas are useful.

Lemma: For each operator P, $\mu(NP) = \mu(PN) = \mu(N)$, where N is the negation operator.

Proof.
$$\mu(PN) = \sup\left\{\frac{||PNx - PNy||}{||x-y||}, \quad x, y \in X,\ x \neq y\right\}$$
$$= \sup\left\{\frac{||P(-x) - P(-y)||}{||-x-(-y)||}, \quad -x, -y \in X,\ -x \neq -y\right\} = \mu(P),$$
$$\mu(NP) = \sup\left\{\frac{||NPx - NPy||}{||x-y||}, \quad x, y \in X,\ x \neq y\right\}$$
$$= \mu(P), \text{ since } ||-Px + Py|| = ||Px - Py||.$$
□

Lemma. For any two operators AB and B on X, $\mu(AB) \leq \mu(A)\mu(B)$.

Proof.
$$\mu(AB) \leq \sup\left\{\frac{||ABx - ABy||}{||x-y||}, \quad x, y \in X,\ x \neq y\right\}$$
$$\leq \sup\left\{\frac{\mu(A)\mu(B)||x-y||}{||x-y||}, x, y \in X,\ x \neq y\right\} = \mu(A)\mu(B),$$

since $||ABx - ABy|| \leq \mu(A)||Bx - By|| \leq \mu(A)\mu(B)$, as $\mu(A)$ and $\mu(B)$ are Lipschitz constants. □

Theorem 5.1. $\mu_p(\bigvee^n) = \mu_p(\bigwedge^n) = (n+1)^{\frac{1}{p}}, \quad$ for $p \geq 1$.

Proof. $|\bigvee^n x_i - \bigvee^n y_i| = |\max\{x_i, x_{i+1}, \ldots, x_{i+n}\} - \max\{y_i, y_{i+1}, \ldots, y_{i+n}\}|$. (Without loss of generality we may assume all the elements concerned to be positive.) Assuming $\bigvee^n x_i - \bigvee^n y_i > 0$ it follows that for some $k \in \{i, i+1, \ldots, i+n\}$

$$\begin{aligned}|\bigvee^n x_i - \bigvee^n y_i|^P &= (x_k - \max\{y_i, y_{i+1}, \ldots, y_{i+n}\})^p \\ &\leq |x_k - y_k|^p, \leq \sum_{j=i}^{i+n} |x_j - y_j|^p.\end{aligned}$$

But then

$$\begin{aligned}||\bigvee^n x - \bigvee^n y||_p^p &\leq \sum_{i=-\infty}^{\infty} \left(\sum_{j=i}^{i+n} |x_j - y_j|^p \right) \\ &\leq (n+1) \sum_{k=-\infty}^{\infty} |x_k - y_k|^p \leq (n+1)||x-y||_p^p.\end{aligned}$$

Thus $|| \bigvee^n x - \bigvee^n y||_p \leq (n+1)^{\frac{1}{p}}$.

To prove that this inequality is sharp, it is sufficient to consider the sequences $x = \langle x_i = \delta_{ij} \rangle$ and $y = 0$. Then

$$||x-y||_p = ||x||_p = 1, \quad \text{and} \quad ||\bigvee^n x - \bigvee^n y||_p = ||\bigvee^n x||_p = (n+1)^{\frac{1}{p}}.$$

The other half follows, since $\bigvee^n = N \bigvee^n N$, by a previous lemma. □

Corollary. $\mu_p(\bigvee^n \bigwedge^n) = \mu_p(\bigvee^n \bigwedge^n) \leq (n+1)^{\frac{2}{p}}$, by the lemma on compositions.

The above upper bound is not sharp, as becomes clear from the following theorem.

Theorem 5.2. For each $x, y \in X$ and $p \geq 1$,

$$||Lnx - Lny||_p \leq (2n+1)^{\frac{1}{p}} ||x-y||_p.$$

Proof. Consider $|Lnx_i - Lny_i|$. Without loss of generality it can be assumed that $Lnx_i \geq Lny_i$, and since Ln is local, all the x_j and y_j concerned can be assumed positive, since adding an arbitrary constant sequence does not change the inequality. Since $Lnx_i = \min\{x_{k-n}, \ldots, x_k\}$, for some $k \in [i, i+n]$, and $Lny_i \geq \min\{y_{k-n}, \ldots, y_k\}$ as well, (it is the maximum of such sets),

$$\begin{aligned}Lnx_i - Lny_i &\leq \min\{x_{k-n}, \ldots, x_k\} - \min\{y_{k-n}, \ldots, y_k\} \\ &\leq \min\{x_{k-n}, \ldots, x_k\} - y_\ell,\end{aligned}$$

for some $\ell \in [k-n, k]$.

But then $Lnx_i - Lny_i \leq x_\ell - y_\ell$, so that $|Lnx_i - Lny_i|^p \leq |x_\ell - y_\ell|^p$, for some $\ell \in [i-n, i+n]$.

Since

$$|x_\ell - y_\ell|^p \le \sum_{\ell=i-n}^{i+n} |x_\ell - y_\ell|^p$$

it follows that

$$||Lnx - Lny||_p^p \le \sum_{i=-\infty}^{\infty} \left(\sum_{\ell=i=n}^{i+n} |x_\ell - y_\ell|^p \right) \le (2n+1) \sum_{i=-\infty}^{\infty} |x_i - y_i|^p$$
$$\le (2n+1)||x-y||_p^p.$$

Taking the pth root on each side proves the theorem. □

Theorem 5.3. For each $p \ge 1$, $\mu_p(Ln) = \mu_p(U_n) = (2n+1)^{\frac{1}{p}}$.

Proof. By the previous theorem it follows that $\mu_p(Ln) \le (2n+1)^{\frac{1}{p}}$. To prove that this is an equality as well, it is sufficient to consider

$$x = \left\langle x_i = \sum_{\ell=j-n}^{j-1} \delta_{\ell i} + \sum_{\ell=j+1}^{j+n} \delta_{\ell j} \right\rangle \quad \text{and} \quad y = \left\langle y_i = \sum_{\ell=j-n}^{j+n} \delta_{0j} \right\rangle.$$

But then

$$||x-y||_p = 1 \quad \text{and} \quad Lny = y,$$

where as $Lnx = 0$ since $\min\{x_i, x_{i+1}, \ldots, x_{i+n}\} = 0$, $\forall i$. Therefore

$$||Lny - Lnx||_p = ||Lny||_p = (2n+1)^{\frac{1}{p}}, \quad \text{and} \quad \mu_p(Ln) \ge (2n+1)^{\frac{1}{p}}.$$

By a previous lemma the dual yields the same minlip constant, proving the theorem. □

Corollary. $\mu_p(LnUn) = \mu_p(UnLn) \le (2n+1)^{\frac{2}{p}}$.

In all the above cases $p = \infty$ can be taken as the limiting case when p tends to ∞. It is easily confirmed that the following theorem holds.

Theorem 5.4. Let S be a syntone selector. Then $\mu_\infty(S) = 1$.

Proof. It is clear that $-||x-y||_\infty \le x_i - y_i \le ||x-y||_\infty$. Let $c = ||x-y||_\infty$. Then, since S is syntone, $y_i - c \le x_i \le y_i + c$ implies that

$$S(y_i - c) \le Sx_i \le S(y_i + c).$$

Since a selector is a smoother, it follows that

$$Sy_i - c \le Sx_i \le Sy_i + c \quad \text{and} \quad -c \le |Sx_i - Sy_i| \le c.$$

This means that $||Sx - S||_\infty \le ||x-y||_\infty$ and that $\mu_\infty(S) \le 1$. A simple example with a constant sequence suffices to show that $\mu_\infty(S) = 1$. □

The last theorem suggests that possibly the previous upper bound for $\mu_p(LnUn)$ is pessimistic, at least for the case where p is large. This can be pursued further, but since the main purpose of a sharp bound for μ_p is the comparison with the "best" or other operators, and investigating the proof of the Lebesgue-type inequality, it is clear that the following restriction to the more general inequality will suffice.

Theorem 5.5. For $x \in X$ and $y \in \mathcal{M}_n$,

$$||LnUnx - LnUny||_p \leq (2n+1)^{\frac{1}{p}} ||x-y||_p$$

and

$$||UnLnx - UnLny||_p \leq (2n+1)^{\frac{1}{p}} ||x-y||_p.$$

Proof. Consider $LnUnx$ and $y \in \mathcal{M}_n$. Assume both $LnUnx$ and $UnLny = y$ are positive. (This can be done without loss of generality since an arbitrary constant sequence can be added to x and y.)

Suppose $LUx_i \geq y_i$. Then $|LUx_i - y_i| \leq |Ux_i - y_i|$, since $Ux_i \geq LUx_i \geq y_i$. If $LUx_i < y_i$, then $|LUx_i - y_i| \leq |Lx_i - y_i|$, since $Lx_i \leq LUx_i < y_i$. But

$$|Ux_i - y_i|,\ |Lx_i - y_i| \leq \max_{j \in [i-n, i+n]} |x_j - y_j|,$$

and therefore

$$|LUx_i - y_i|^p \leq \max_{j \in [i-n, i+n]} |x_j - y_j|^p \leq \sum_{j=i-n}^{i+n} |x_j - y_j|^p,$$

$$\begin{aligned} ||LUx - y||_p^p &\leq \sum_{i=-\infty}^{\infty} \left(\sum_{j=i-n}^{i+n} |x_j - y_j|^p \right) = (2n+1) \sum_{j=-\infty}^{\infty} |x_j - y_j|^p \\ &= (2n+1||x-y||_p^p. \end{aligned}$$

This implies that $||LUx - y||_p \leq (2n+1)^{\frac{1}{p}} ||x-y||_p$. □

This inequality is sharp, as the example used previously, with the sequences

$$x = \left\langle x_i = \sum_{\ell=j-n}^{j-1} \delta_{\ell_i} + \sum_{\ell=j+1}^{j+n} \delta_{\ell i} \right\rangle \quad \text{and} \quad y = \left\langle y_i = \sum_{\ell=j-n}^{j+n} \delta_{ij} \right\rangle,$$

demonstrates that

$$||Lny - Lnx||_p = ||UnLny - 0||_p \geq (2n+1)^{\frac{1}{p}} \quad \text{but} \quad ||x-y||_p = 1.$$

At this stage, for the operators $UnLn$ and $LnUn$, a Lebesgue type Comparison Theorem can be formulated for the set $\mathcal{M}_n$ of n-monotone sequences.

Theorem 5.6. Let P be either $UnLn$ or $LnUn$. Then, for each $s \in \mathcal{M}_n$, $||Px-x||_p \leq (1+(2n+1)^{\frac{1}{p}})||s-x||_p$.

Proof.
$$\begin{aligned} ||Px-x||_p &\leq ||Px-s||_p + ||s-x||_p, \ \forall s \in \mathcal{M}_n \\ &\leq (2n+1)^{\frac{1}{p}}||s-x||_p + ||s-x||_p \\ &\leq (1+(2n+1)^{\frac{1}{p}})||s-x||_p. \end{aligned}$$
□

If there is a "best" approximation to x in $\mathcal{M}_n$ then clearly this inequality reassures that $LnUn$ and $UnLn$ yield approximations that are "less than best" by the factor $1+(2n+1)^{\frac{1}{p}}$. Whether this factor is optimal is not established yet.

The above theorem can be generalized easily to facilitate the comparison of more general smoothers with the best approximation in $\mathcal{M}_n$. This can be done with the following theorem.

Theorem 5.7. Let $x \in X$ and $y \in \mathcal{M}_n$. If $P \in [L_n, U_n]$, then $||Px-y||_p \leq (2n+1)^{\frac{1}{p}}||x-y||_p$.

Proof. Consider the ith elements of x and y. (Assume both are positive. If not, a constant can be added temporarily for the neighborhood of i under consideration.)

If $y_i > P_n x_i$, then $|y_i - Px_i| \leq |y_i - L_n x_i|$. If $y_i \leq Px_i$, then $|Px_i - y_i| \leq |U_n x_i - y_i|$. Since $L_n y = U_n y = y$ it follows that $|Px_i - y_i|^p \leq \max\{|U_n x_i - U_n y_i|^p, |L_n x_i - L_n y_i|^p\}$.

By the argument used in the proof of Theorem 5.2, it follows that

$$|Px_i - y_i|^p \leq \sum_{\ell=i-n}^{i+n} |x_\ell - y_\ell|^p, \quad \text{or} \quad ||Px-y||_p^p \leq (2n+1)^{\frac{1}{p}}||x-y||_p^p.$$

This proves the theorem. □

There are several popular operators that meet the requirements of the above theorem, such as medians and their powers. For example a simple popular smoother like $M_2M_1^2$ can be compared with the best approximation in $\mathcal{M}_2$ by noting that $L_2U_2L_1U_1$ is in the interval $[L_2, U_2]$, since $L_2U_2L_1U_1 \leq L_2U_2U_1 \leq L_2U_2U_2 = L_2U_2 \leq U_2$ etc. Others are the operators $\frac{1}{2}(UL+LU)$, $U+L-I$, etc.

It is almost natural, given the results of previous comparisons, to guess that the "best approximations" are in the corresponding $LULU$-interval. This turns out to be not true in general, as the following example will demonstrate.

Example. Let $x = \langle x_i = \delta_{i,j} + \delta_{i,1+j}\rangle$ be a sequence, that clearly has two consecutive impulses superimposed on the zero sequence.

With $L_2U_2x = U_2L_2x = 0$, it is clear that 0 is not a best approximation in any norm with $p < \infty$ since $y = \langle y_i = \delta_{i,j-1} + \delta_{i,j} + \delta_{i,j+1}\rangle$ is better.

It is worth noting that a best approximation from $\mathcal{M}_2$ does not even lie between L_2x and U_2x, and does not remove the basic impulse for which U_2L_2, L_2U_2, M_2 etc. were designed! (The possible exception is the $p = \infty$ norm, where a best approximation may be in the $LULU$-interval.) The above example illustrates this since y is a best approximation as $||x - y||_p = 1$ cannot be improved.

This casts doubt on the basic idea of approximating with $\mathcal{M}_n$ as a general method of impulsive noise removal, and highlights the remarkable ability of the $LULU$-smoothers and similar local smoothers. The indications are however that for $n = 1$ the "best approximation" from M_1 may well be in the $LULU$-interval. This needs further investigation, but does not seem very important.

An alternative, secondary approximation problem presents itself naturally. Given a choice of a smoother S amongst the smoothers in a class $\mathcal{C}$ that map into the range of S, and being aware of the fact that a smoother can at best destroy information selectively, the question of comparison with alternatives comes to mind. Is there an associated smoother that can somehow reduce the damage done by S? If so, is there a smoother S_L such that the residual $(S_LS - I)x$ is minimal, for each sequence x, in some relevant norm? Similarly, is there a smoother S_R such that $(SS_R - I)x$ is minimal? A simple example is sufficient to show that S_R does not exist in respect of, for example, the 1-norm.

With respect to the Total Variation, which is a norm in ℓ_1, there always is a left quasi-inverse. Indications are also that a left quasi-inverse S_L exists, with respect to the 1-norm. Thus $\|(S_LS - I)x\|_1 \leq \|(AS - I)x\|_1$, for each sequence $x \in \ell_1$ and each composition A of the basic operators $\vee$ and $\wedge$. In particular each idempotent composition, (which includes all $LULU$-operators) has the identity operator as quasi-inverse. These operators are thus best approximations to the identity operator, with respect to left-composition. Articles investigating these issues, and interesting incidental results may appear soon.

6. Variation Reduction and Shape Preservation

Ich Forscher – ich? Oh spart dies Wort! –
Ich bin nur schwer – so manche Pfund!
Ich falle, falle immerfort
Und endlich auf den Grund

Nietzsche

There are a variety of concepts of smoothness in functions; Continuity, Differentiability, Integrability, Measurability, to name but a few. None of these generalize suitably to sequences. There are other concepts that are linked to these (27), two of which are relevant here; Bounded Variation and (Piece-wise-) Monotonicity.

Two examples that are well known can be taken to illustrate.

1) If f is an increasing real-valued function on $[a, b]$, then f is differentiable almost everywhere and $\int_a^b f'(x) \leq f(b) - f(a)$.
2) A function f is of bounded variation on $[a, b]$ if and only if f is the difference of two monotone real-valued functions.

Clearly piecewise monotonicity is sufficient in both.

The digital world has impressed us with the reality that, in practice, measurements are sequences. So are transmitted signals, yet the concepts of smoothness often used are based on concepts of smoothness of an "underlying" function or "true function". For the Fast Fourier Transform (FFT) and Fast Wavelet Transform the underlying assumption of a band-limited function is usual. There is something unsatisfactory in basing the concept of smoothing used in the subject of Nonlinear Smoothing on properties of an underlying function which is not at hand. Similarly concepts like "trend" and "impulse" are often also based by implication on linear functions and delta functions. Most people using median smoothers feel that there are some very useful properties that are "almost always" there, the behavior is "enigmatic" and unpredictable. Yet they generally behave very well. If measures and concepts of smoothness that are more directly attributable to a given sequence are to be used, it seems natural to turn to variation and monotonicity. When these concepts are chosen appropriately they turn out to be illuminating and reveal a very strong underlying mathematical framework for the analysis of signals, even with a wide variety of contamination.

Both concepts can be used globally and locally. For practical purposes of measurements sequences are always finite, but there is no problem generalizing to $X = \ell_p$ and a natural choice is $p = 1$.

We have already become used to local monotonicity as a local concept of smoothness (there are no pulses of lesser briefness), and a connection to variation can be envisaged as follows.

Definition. $Tx = T(x) = ||\triangle x||_1$, where $\triangle$ is the (forward-) difference operator. Tx is called the (total-) variation of x.

It is clear that if $x \in \ell_1$, then $\triangle x \in \ell_1$, since

$$\sum_{i=-N}^{N} |x_{i+1} - x_i| \leq \sum_{i=-N}^{N} (|x_i| + |x_{i+1}|) \leq 2||x||_1.$$

Theorem 6.1: (i) $T(Ex) = T(x)$, where E is the shift operator.
(ii) $T(x + y) \leq T(x) + T(y)$ (subadditivity).
(iii) $T(\alpha x) = |\alpha| T(x)$ and $T(x) = 0$ only if $x = 0$.

Proof.

$$\text{(i)} \qquad \sum_{i=-N}^{N} |Ex_{i+1} - Ex_i| = \sum_{i=-N}^{N} |x_{i+2} - x_{i+1}| = \sum_{i=-N+1}^{N+1} |x_{i+1} - x_i|.$$

Taking limits proves that $T(Ex) = T(x)$.
(ii) $|(x + y)_{i+1} - (x + y)_i| = |x_{i+1} - x_i + y_{i+1} - y_i| \leq |x_{i+1} - x_i| + |y_{i+1} - y_i|$.
(iii) $|(\alpha x)_{i+1} - (\alpha x)_i| \leq |\alpha|\, |x_{i+1} - x_i|$.
Summing and taking limits proves $T(x+y) \leq T(x)+T(y)$ and $T(\alpha x) = \alpha T(x)$. □

Since a sequence that is n-monotone is $(n-1)$-monotone, $\mathcal{M}_0 \supset M_1 \supset M_2 \supset \cdots \supset M_n \supset \cdots$ is a nested sequence of subsets of ℓ_1 whose intersection contains only the zero sequence. Intuitively it is clear that the progressive mapping of a sequence $x \in \ell_1$ into these subsets should reduce the variation, ultimately arbitrarily close to 0. There must be a connection between local monotonicity and total variation. Since it is possible to decompose the operators $U_n L_n$, $L_n U_n$, L_n and U_n into compositions of two basic local maximum and minimum operators, by $U_n = (\bigwedge^n \bigvee^n)$ and $L_n = \bigvee^n \bigwedge^n$, the following approach is logical.

For convenience U_1 can be denoted by U and L_1 by L provisionally.

Lemma. If j is a point where $Ux_j \neq x_j$, then $Ux_j - x_j = \min\{x_{j-1}, x_{j+1}\} - x_j > 0$ and $Ux_{j+1} = x_{j+1}$.

Proof. Let $Ux_j \neq x_j$. Then $Ux_j > x_j$, and by definition

$$\begin{aligned}(Ux)_j &= \min\{\max\{x_{j-1}, x_j\}, \max\{x_j, x_{j+1}\}\} \\ &= \min\{x_{j-1}, x_{j+1}\},\end{aligned}$$

since both maxima must be larger than x_j. Noting that

$$\begin{aligned}(Ux)_{j+1} &= \min\{\max\{x_j, x_{j+1}\}, \max\{x_{j+1}, x_{j+2}\}\} \\ &= \min\{x_{j+1}, \max\{x_{j+1}, x_{j+2}\}\} \le x_{j+1},\end{aligned}$$

it follows that $Ux_{j+1} = x_{j+1}$. □

It is clear that Ux cannot differ from x at consecutive points and that they differ at precisely a point where the value is smaller than both neighbors.

Theorem 6.2. $T(x) = T(\bigvee x) + 2||Ux - x||_1$ for all $x \in \ell_1$.

Proof. Consider the sequence $t = \{t_j\}$ of integers where $Ux_j > x_j$. Then $T(x) = \sum_{j=-\infty}^{\infty} \tau_j$ where $\tau_j = \sum_{i=t_j-1}^{t_{j+1}-2} |x_{i+1} - x_i|$, simply partitions the sum.

Consider a specific j, and letting $t_j = k$ and $t_{j+1} = m$, let n be the first index after k where $x_n > x_{n+1}$. Since $x_{k+1} > x_k$ and $x_m < x_{m-1}$, n must exist and be between k and m, and $x_k, x_{k+1}, \ldots, x_n$ is non-decreasing and $x_n, x_{n+1}, \ldots, x_m$ non-increasing, and

$$\begin{aligned}\mu_j &= \sum_{i=k-1}^{n-1} |\bigvee x_{i+1} - \bigvee x_i| + \sum_{i=n}^{m-2} |\bigvee x_{i+1} - \bigvee x_i| \\ &= |\bigvee x_k - \bigvee x_{k-1}| + \sum_{i=k}^{n-2} |x_{i+2} - x_{i+1}| + \sum_{i=n}^{m-2} |x_{i+1} - x_i|,\end{aligned}$$

since

$$\bigvee x_n - \bigvee x_{n-1} = 0.$$

Except for three terms this yields T_j and therefore;

$$\begin{aligned}\tau_j &= \mu_j - |x_{k+1} - x_{k-1}| + |x_k - x_{k-1}| + |x_{k+1} - x_k| \\ &= \mu_j - |x_{k+1} - x_{k-1}| + x_{k-1} - x_k + x_{k+1} - x_k \\ &= \mu_j - \max\{x_{k+1}, x_{k-1}\} + \min\{x_{k+1}, x_{k-1}\} + x_{k-1} + x_{k+1} - 2x_k \\ &= \mu_j + 2(\min\{x_{k+1}, x_{k-1}\} - x_k) \quad \text{(From the previous lemma)} \\ &= \mu_j + 2(Ux_k - x_k)\end{aligned}$$

Therefore

$$\begin{aligned}T(\bigvee x) &= \sum_{j=-\infty}^{\infty} \mu_j = \sum_{j=-\infty}^{\infty} \tau_j - 2 \sum_{j=-\infty}^{\infty} |Ux_{t_j} - x_{t_j}| \\ &= T(x) - 2 \sum_{i=-\infty}^{\infty} |Ux_i - x_i| = T(x) - 2||Ux - x||_1.\end{aligned}$$

□

It is clear that the operator $\bigvee$ is variation decreasing in a precise sense. A similar property must exist for $\bigwedge$, and can be demonstrated by the following corollary to Theorem 6.2.

Corollary. $T(x) = T(\bigwedge x) + 2||Lx - x||_1$.

Proof. $T(\bigwedge x) = T(E \bigwedge x) = T(\bigwedge(Ex)) = T(-\bigwedge(Ey))$, and with $y = -x$,

$$\begin{aligned} T(\bigwedge x) &= T(\bigvee y) = T(y) - 2||Uy - y||_1 \\ &= T(-Ex) - 2||U(-Ex) + Ex||_1 \\ &= T(Ex) - 2|| - L(Ex) + Ex||_1 \\ &= T(x) - 2||E(x - Lx)||_1 \end{aligned}$$

$$= T(x) - 2||x - Lx||_1,$$

which proves the corollary. □

Corollary. Any operator O that is a composition of $\bigvee$ and $\bigwedge$ is variation diminishing (variation non-increasing).

Proof. For $\bigvee$ and $\bigwedge$ the statement follows from Theorem 6.2 and its corollary, since $T(\bigwedge x)$ and $T(\bigvee x)$ exceed $T(x)$ by amounts $2||Lx - x||_1$ and $2||Ux - x||_1$ respectively. The rest of the proof follows by induction. □

The following theorem shows that $\bigwedge$ does not reduce variation on any output of $\bigvee$.

Theorem 6.3. $T(\bigwedge\bigvee x) = T(\bigvee x)$ and $T(\bigvee\bigwedge x) = T(\bigwedge x)$ for each $x \in \ell_1$.

Proof. Suppose that $U(\bigwedge x)_i > \bigwedge x_i$ for some index i. Then $(\bigwedge x)_{i-1} > (\bigwedge x)_i < (\bigwedge x)_{i+1}$. If $(\bigwedge x)_i$ is smaller than $\min\{x_{i-2}, x_{i-1}\}$ and $\min\{x_i, x_{i+1}\}$ it follows that it is smaller than x_{i-1} and x_i. This is a contradiction since $(\bigwedge x)_i$ is the smaller of x_{i-1} and x_i. Thus $U(\bigwedge x) = \bigwedge x$. But then

$$\begin{aligned} T(\bigwedge x) &= T(\bigvee\bigwedge x) + 2||U(\bigwedge x) - \bigwedge x||_1 \\ &= T(\bigvee\bigwedge x). \end{aligned}$$

A similar proof can be given for the other part of the theorem, or the usual duality argument can be used; If $y = -x$, then

$$\begin{aligned} T(\bigvee x) &= T(-\bigwedge y) = T(\bigwedge y) = T(\bigvee\bigwedge y) \\ &= T(-\bigwedge\bigvee x) = T(\bigwedge\bigvee x). \end{aligned}$$ □

Corollary. $T(\bigvee Ux) = T(Ux)$ and $T(\bigwedge Lx) = T(Lx)$.

Example 1. Let

$$x_i = \begin{cases} 1 & \text{for } i = -1, 1 \\ 0 & \text{elsewhere} \end{cases}.$$

Then

$$Ux_i = \begin{cases} 1 & \text{for } i = -1, 0, 1 \\ 0 & \text{elsewhere} \end{cases} \quad \text{and} \quad \bigvee x = \begin{cases} 1 & \text{for } i = -2, -1, 0, 1 \\ 0 & \text{elsewhere} \end{cases}.$$

Elementary computation yields

$$T(x) = 4, \quad T(Ux) = 2, \quad T(\bigvee x) = 2 \quad \text{and} \quad ||Ux - x||_1 = 1.$$

Example 2. Let

$$x_i = \frac{(-1)^i}{i^2}, \quad i \neq 0 \text{ and } x_0 = 1.$$

$$T(x) = 2\left(2 + \sum_{i=1}^{\infty}\left(\frac{1}{i^2} + \frac{1}{(i+1)^2}\right)\right) = 4 + 2\sum_{i=1}^{\infty}\frac{1}{i^2} + 2\sum_{i=1}^{\infty}\frac{1}{(i+1)^2}$$
$$= 2 + 2\,\frac{\pi^2}{6} + 2\left(\frac{\pi^2}{6} - 1\right) = \frac{2}{3}\pi^2 + 2$$

and

$$T(Ux) = T(\bigvee x) = 2,$$

since $\bigwedge x$ is monotone increasing up to -1 and monotone decreasing from 0. Elementary calculation gives

$$||Ux - x|| = \frac{\pi^2}{3}.$$

It should be clear, since $U = U_1 = \bigwedge\bigvee$ and $L = L_1 = \bigvee\bigwedge$, that the following theorem should hold, because of the fact that points where $U_1 x_i \neq x_i$ are separated by at least one index.

Theorem 6.4. $T(x) = T(Ux) + T(Ux - x)$ and $T(x) = T(Lx) + T(Lx - x)$.

Proof. Let t_i be the sequence of points such that $Ux_{t_j} > x_{t_j}$. By the first lemma such a point x_{t_j} has neighbors x_ℓ in x_r such that $Ux_\ell - x_\ell = 0$ and $Ux_r - x_r = 0$. Therefore, by Theorem 6.1,

$$\begin{aligned} T(Ux - x) &= \sum_{j=-\infty}^{\infty} |Ux_{t_j} - x_{t_j} - (Ux_\ell - x_\ell)| + |Ux_{t_j} - x_{t_j} - (Ux_r - x_r)| \\ &= \sum_{j=-\infty}^{\infty} 2|Ux_{t_j} - x_{t_j}| = 2||Ux - x||_1, \end{aligned}$$

$$\begin{aligned} T(x) &= T(\bigvee x) + 2||Ux - x||_1 \\ &= T(Ux) + 2||Ux - x||_1, \quad \text{by Theorem 6.3,} \\ &= T(Ux) + T(Ux - x). \end{aligned}$$

The other inequality follows similarly, or by a duality argument, as before. □

The idempotent operators U and L do not have a common range, but the range of LU and UL is common and is $\mathcal{M}_1$. The precise variation reduction relations for the composite operators are surprisingly simple to prove. This yields a quantitative measure for the contribution to the total variation of x by single downward pulses and single upward pulses, as interpreted by LU and UL respectively.

Theorem 6.5. $T(x) = T(LUx) + T(LUx - x)$ and $T(x) = T(ULx) + T(ULx - x)$.

Proof. By Theorem 6.1, $T(x) \leq T(LUx) + T(LUx - x)$, by the triangle inequality. By Theorem 6.4,

$$\begin{aligned} T(x) &= T(Ux) + T(Ux - x) \\ &= T(LUx) + T(LUx - Ux) + T(Ux - x) \\ &\geq T(LUx) + T(LUx - x), \quad \text{by the subadditivity.} \end{aligned}$$

This proves the first half of the theorem and the second is proved similarly, or by the usual duality argument. □

Corollary.

$$\begin{aligned} T(LUx - x) &= T(LUx - Ux) + T(Ux - x) \\ &= 2\|LUx - Ux\|_1 + 2\|Ux - x\|_1, \\ T(ULx - x) &= T(ULx - Lx) + T(Lx - x) \\ &= 2\|ULx - Lx\|_1 + 2\|Lx - x\|_1. \end{aligned}$$

For comparison it can be shown that for the median operator M_1 the decrease in variation can be arbitrarily slow.

Example 3. Let

$$x = \begin{cases} (-1)^i & \text{for } i \in [-n, n], \\ 0 & \text{elsewhere} \end{cases}.$$

Then

$$T(x) = 4\frac{n(n+1)}{2} + 2 \quad \text{and} \quad M_1 x = \begin{cases} -(-1)^i & \text{for } i \in [-n+1, n-1], \\ 0 & \text{elsewhere} \end{cases}.$$

Therefore $T(M_1 x) = 2(n-1)n + 2$ and the relative decrease in variation is

$$\frac{4n}{2n(n+1)+2} < \frac{n}{n(n+1)} < \frac{2}{n}.$$

(For the same example) $T(U_1 x) = T(L_1 U_1 x) = 2$ and $T(L_1 x) = T(U_1 L_1 x) = 2$.

The above example again illustrates the general "enigmatic" behavior of the median smoothers and their compositions, despite the fact that they are so closely related to the $LULU$ operators, and the preference given to their iterates, in spite of the computational effort and generally marginal changes produced by repetition.

At this stage a generalization of the previous theorem to general U_nL_n, L_nU_n and U_nL_n is possible in several ways, but the proofs become difficult, if pursued along similar lines. There is a surprising shortcut, yielding in the process some new results which are very surprising. As it turns out there are stronger trend preserving properties , and variation preserving properties, of the *LULU*-operators. The median smoothers M_n are trend preserving, in the sense that if $x_{i-n}, \ldots, x_i, \ldots, x_{i+n}$ is monotone, then $(M_nx)_i = x_i$. The operators L_nU_n and U_nL_n share this property.

Another, stronger, result is the global sequence preservation $L_nU_nx = U_nL_nx = x$ if x is n-monotone. The operator M_n shares this property. This is however a strong requirement, which can be weakened if only local order preservation is required. There is a different, more local, and strong trend preservation, which the medians do not share.

Definition. An operator A is called neighbor trend preserving (*ntp*) if, for each sequence $x \in X; x_{i+1} \leq x_i \Rightarrow Ax_{i+1} \leq Ax_i$, and $x_{i+1} \geq x_i \Rightarrow Ax_{i+1} \geq Ax_i$ for each index i.

Theorem 6.6. The operators $\bigvee\bigwedge$ and $\bigwedge\bigvee$, as well as all compositions of them, are neighbor trend preserving.

Proof. Suppose that $x_{i+1} \geq x_i$. Then

$$(\bigvee x)_i = \max\{x_{i+1}, x_i\} = x_{i+1},$$

and

$$(\bigvee x)_{i+1} = \max\{x_{i+2}, x_{i+1}\} \geq x_{i+1} = (\bigvee x)_i.$$

But then

$$(\bigwedge\bigvee x)_{i+1} = \min\{\bigvee x_{i-1}, \bigvee x_i\} = \bigvee x_i$$

and

$$(\bigwedge\bigvee x)_i = \min\{\bigvee x_i, \bigvee x_{i-1}\} \leq \bigvee x_i = (\bigwedge\bigvee x)_{i+1}.$$

A similar argument holds for the case when $x_{i+1} \leq x_i$, proving that $\bigwedge\bigvee$ is trend preserving. Similarly, or for some variation, a typical duality argument can be used to prove the other half.

Considering the sequence x and its negative $-x$;

$$x_{i+1} \gtreqless x_i \Leftrightarrow (-x)_{i+1} \lesseqgtr (-x)_i.$$

Since $\bigwedge\bigvee$ is trend preserving,

$$(\bigwedge\bigvee(-x))_{i+1} \lesseqgtr (\bigwedge\bigvee(-x))_i.$$

But $\bigvee(-x) = -\bigwedge x$ and $(\bigwedge(-x) = -\bigvee(x)$ so that

$$(-\bigvee\bigwedge x)_{i+1} \lesseqgtr (-\bigvee\bigwedge x)_i \quad \text{or} \quad (\bigvee\bigwedge x)_{i+1} \gtreqless (\bigvee\bigwedge x)_i. \qquad \square$$

The above theorem is strong but not very general, since the only compositions of the operators $U_1 = \bigvee\bigwedge$ and $L_1 = \bigwedge\bigvee$ are U_1L_1, and L_1U_1. (The operators $\bigwedge$ and $\bigvee$ themselves are not *ntp*, as a simple example can prove: $x_0 = 1, x_1 = -1, x_2 = 2$, but $(\bigvee x)_0 = 1, (\bigvee x)_1 = 2)$.

The important general *LULU*-operators L_n and U_n are however *ntp*, as the following theorem proves.

Theorem 6.7. U_n and L_n are neighbor trend preserving.

Proof. Suppose $x_{i+1} \geq x_i$ and $U_nx_{i+1} < U_nx_i$. For notational convenience i can be assumed to be 0.

$$\begin{aligned} U_nx_0 &= \min\{\max\{x_{-n}, \ldots, x_0\}, \ldots, \max\{x_0, \ldots, x_n\}\} \\ &\leq \min\{\max\{x_{-n+1}, \ldots, x_0, x_1\}, \ldots, \max\{x_0, \ldots, x_n\}\}, \\ U_nx_1 &= \min\{\max\{x_{-n+1}, \ldots, x_0, x_1\}, \ldots, \max\{x_1, \ldots, x_{n+1}\}\}. \end{aligned}$$

If the latter minimum is smaller than U_nx_0 it means that it must be because $\max\{x_1, \ldots, x_{n+1}\}$ is smaller than all the others involved. In particular

$$\max\{x_1, \ldots, x_{n+1}\} < \max\{x_0, \ldots, x_n\}.$$

This can only be true if $x_0 > x_1, \ldots, x_{n+1}$, and in particular $x_0 > x_1$ is a contradiction. Thus $U_nx_1 \geq U_nx_0$.

A similar argument shows that $x_i \geq x_{i+1}$ implies $U_nx_i \geq U_nx_{i+1}$. As in the previous proof a simple duality argument proves the other half of the theorem. □

What is surprising is the weak demand for $(U_nx)_i \lesseqgtr (U_nx)_{i+1}$ to hold. This demand is independent of n! It only requires $x_i \lesseqgtr x_{i+1}$. One of the consequences is a simple proof of a surprising general theorem, which is simple to envisage and prove, but which would not have been easy to see as significant, due to the strong assumptions required for it to hold. Since these strong demands are demonstrated to hold for all the operators L_n and U_n, as well as all compositions of them, the significance of the theorem becomes clear.

Definition. An operator S is difference reducing if $|Sx_{i+1} - Sx_i| \leq |x_{i+1} - x_i|$ for each sequence $x \in X$ and each index i.

Definition. An operator S is fully trend preserving if it is ntp and difference reducing.

Theorem 6.8. Let P be a fully trend preserving operator. Then $T(x) = T(Px) + T(x - Px)$ for each sequence $x \in \ell p$.

Proof. $|x_{i+1} - x_i| = |Px_{i+1} - Px_i + x_{i+1} - Px_{i+1} - (x_i - Px_i)|$. If $x_{i+1} \geq x_i$, then $Px_{i+1} - Px_i \geq 0$ and $x_{i+1} - Px_{i+1} - (x_i - Px_i) = (x_{i+1} - x_i) - (Px_{i+1} - Px_i) \geq 0$. Thus $|x_{i+1} - x_i| = |Px_{i+1} - Px_i| + |x_{i+1} - Px_{i+1} - (x_i - Px_i)|$.

If $x_i \leq x_{i+1}$, a similar argument yields the same. Summing both sides gives

$$\sum_{i=-N}^{N} |x_{i+1} - x_i| = \sum_{i=-N}^{N} |Px_{i+1} - Px_i| + \sum_{i=-N}^{N} |(I-P)x_{i+1} - (I-P)x_i|.$$

Taking limits yields the required result. □

Theorem 6.9. U_n and L_n are difference reducing.

Proof. Consider a sequence x and an index i. Without loss of generality, we may assume that $x_{i+1} \geq x_i$ and therefore $Ux_{i+1} \geq Ux_i$. Since equality would yield a trivial result, we assume that $Ux_{i+1} > Ux_i$.

If $Ux_{i+1} = x_{i+1}$, then $0 \leq Ux_{i+1} - Ux_i \leq x_{i+1} - Ux_i \leq x_{i+1} - x_i$, since $Ux_j \geq x_j$, for each index j. Assume therefore that $U_n x_{i+1} \neq x_{i+1}$, so that $U_n x_{i+1} > x_{i+1}$. $U_n x_{i+1} = \min\{\max\{x_{i+1-n}, \ldots, x_{i+1}\}, \ldots, \max\{x_{i+1}, \ldots, x_{i+1+n},\}\} > x_{i+1}$, and each of $\max\{x_{i+1-n}, \ldots, x_{i+1}\}, \ldots, \max\{x_{i+1}, \ldots, x_{i+1+n}\} > Ux_{i+1}$.

Investigating each of the subsets involved in

$$U_n x_i = \min\{\max\{x_{i-n}, \ldots, x_i\}, \ldots, \max\{x_i, \ldots, x_{i+n}\}\},$$

we see that each must have an element not smaller than $U_n x_{i+1}$, and thus $U_n x_i \geq U_n x_{i+1}$. Together with the inequality $U_n x_{i+1} \geq Ux_i$ above, it follows that $U_n x_{i+1} = U_n x_i$. Thus $|U_n x_{i+1} - U_n x_i| \leq |x_{i+1} - x_i|$, for all the cases, and all indexes i.

A similar proof for $L = L_n$ completes the proof. (or the usual duality argument can derive the inequality for L from that of U, since $U(-x) = -L(x)$). □

Corollary.
(a) $T(x) = T(U_n x) + T(x - U_n x)$.
(b) $T(x) = T(L_n x) + T(x - L_n x)$.
(c) $T(x) = T(L_n U_n x) + T(x - L_n U_n x)$.
(d) $T(x) = T(U_n L_n x) + T(x - U_n L_n x)$.

Proof. Each of the operators satisfies the demands of Theorem 6.8. □

It is instructive to illustrate with some simple examples, some direct application of the above results. If a signal is expected in a measurement contaminated by impulsive noise and identically independently distributed noise from a "good" (perhaps near Gaussian-) distribution, then there may be an a priori estimate of the total variation of the signal expected. If s is the sequence of the pure signal, r the well-behaved noise and e the impulsive noise, then the sequence $x = s + r + e$ and generally $T(x) \leq T(s) + T(r) + T(e)$. Assuming that the impulsive sequence has brief large impulses, (its contribution may be large) and that $T(r) \ll T(x)$ (the signal-to-noise ratio is good) we can approximate $T(x) \simeq T(s) + T(e) + T(r)$. Experimentation with the type of random noise, or theoretical estimation, can determine how large $T(r)$ is approximately. Smoothing successively by $L_1U_1, L_2U_2, \ldots, L_nU_n$, yields a monotone reduction of $T(L_iU_i)$. As soon as large reductions in variation stop, it indicates that the impulsive noise has been adequately removed, and provided $T(L_nU_n)$ is near the expected variation of $T(s)$, not too much damage has been incurred. At each level the variation that is removed is calculable, and therefore indicates the contribution by noise at that resolution level. The alternative decomposition with U_nL_n will yield an interval of ambiguity.

Example 4. The *LULU*-decompositions of a sequence of identically, independently generated random noise from a cubic B-spline distribution, show the fast reduction to an expected 0 variation in the smoothed components. (This could be theoretically derived.)

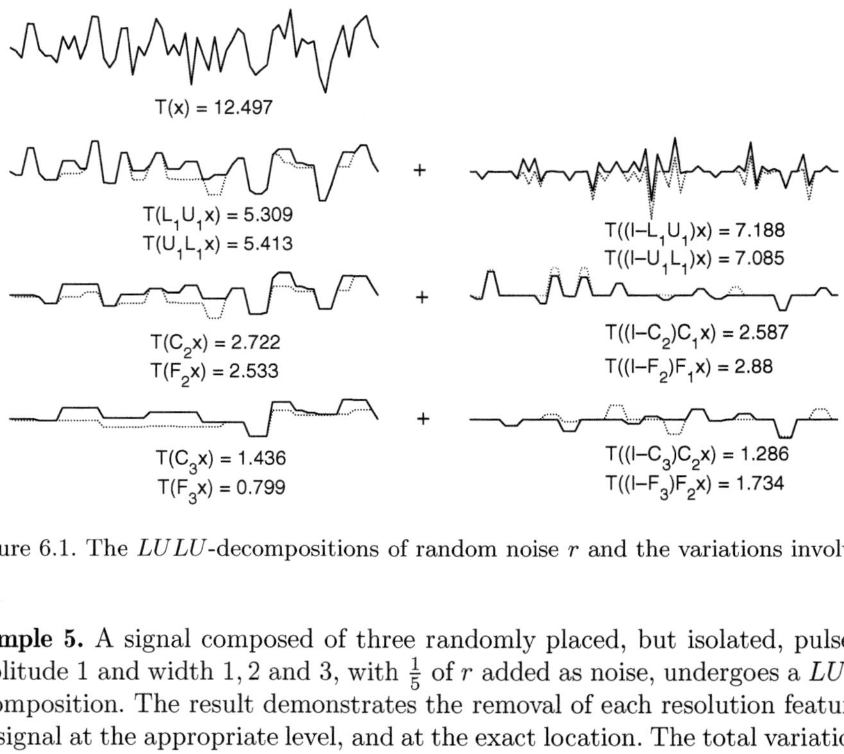

Figure 6.1. The *LULU*-decompositions of random noise r and the variations involved.

Example 5. A signal composed of three randomly placed, but isolated, pulses of amplitude 1 and width 1, 2 and 3, with $\frac{1}{5}$ of r added as noise, undergoes a *LULU*-decomposition. The result demonstrates the removal of each resolution feature of the signal at the appropriate level, and at the exact location. The total variation of the signal is 6, so that the total variation will be roughly that, plus the variation of the noise r. Comparing the contribution expected from r in Example 4, the reduction in variation is predictable.

Example 6. A slowly oscillating sinusoid is considered as the signal and the signal of Example 5 is added as noise. Decomposition yields a fair approximation to the signal, despite the impulses of considerable amplitude. (Multiplying the impulses by 10 or 10^{10} would not significantly change the result, except for different total variations peeled off at each layer.

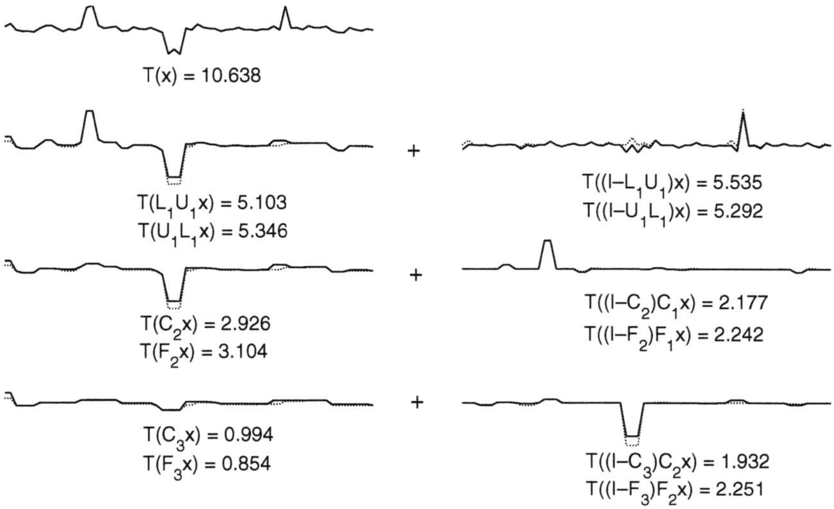

Figure 6.2. The $LULU$-decompositions of 3 pulses with random noise r.

The above simple examples demonstrate why the $LULU$-decompositions have been so practical in applications of measurements contaminated by bad impulsive noise, and well-distributed noise. Noting that the differences between the decompositions of L_nU_n and U_nL_n were very small, it is clear that decomposition with median operators (a highly regarded procedure, in two dimensions – considered much better than wavelet-decomposition in many applications (1)) would yield similar results, but certainly not better. The computational and conceptual simplicity of the $LULU$-smoothers yields a significant advantage. When the two $LULU$-decompositions differ significantly at any level, it implies that there is a significant amount of oscillation at the Nyquist-frequency of that level. The median would behave unpredictably there (Example 3).

Fully trend preserving operators have further remarkable properties with surprising strong results. The following theorems are basic and easy, if somewhat surprising in the last case and the corollary.

Theorem 6.10. Let A and B be ftp on sequences in ℓp. Then:

(a) AB and BA are ftp.

(b) $\alpha A + (1 - \alpha)B$ is ftp for all $\alpha \in [0, 1]$

(c) $I - A$ is ftp.

Proof. (a) Suppose $x_{i+1} \geq x_i$. Then $Ax_{i+1} \geq Ax_i$, because A is ftp and $B(Ax)_{i+1} \geq B(Ax)_i$ because B is ftp. Therefore $(BA)x_{i+1} \geq (BA)x_i$. A similar argument demonstrates that if $x_{i+1} \leq x_i$, then $(BA)x_{i+1} \leq (BA)x_i$. Also $|Ax_{i+1} - Ax_i| \leq |x_{i+1} - x_i|$ and $|B(Ax)_{i+1} - B(Ax)_i| \leq |Ax_{i+1} - Ax_i| \leq |x_{i+1} - x_i|$. This proves that BA is ftp, and a similar argument demonstrates that AB is ftp.

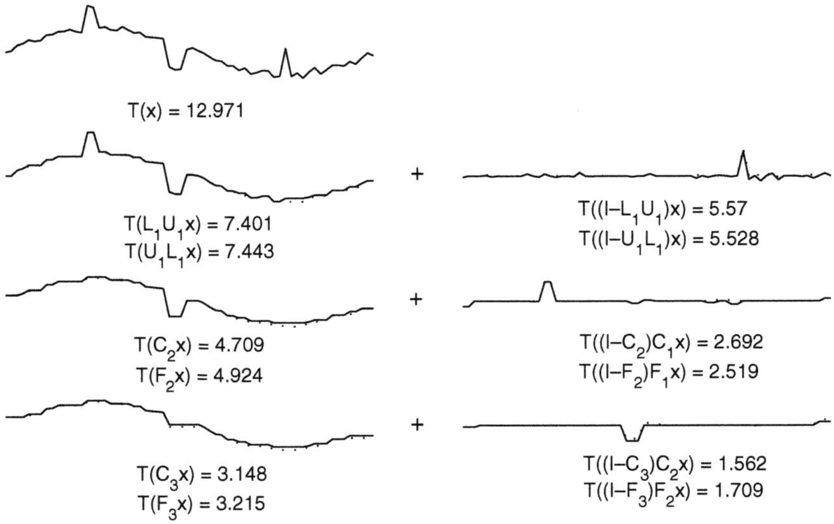

Figure 6.3. The $LULU$-decomposition of a sinusoid, impulsive noise and r.

(b) $|(\alpha A + (1-\alpha)B)x_{i+1} - (\alpha A + (1-\alpha)B)x_i| \leq \alpha|Ax_{i+1} - Ax_i| + (1-\alpha)|Bx_{i+1} - Bx_i| \leq \alpha|x_{i+1} - x_i| + (1-\alpha)|x_{i+1} - x_i| \leq |x_{i+1} - x_i|$. Clearly as A and B are ftp, $(\alpha A + (1-\alpha)B)x_{i+1} \geq (\alpha A + (1-x)B)x_i$ if $x_{i+1} \geq 0$ etc.

(c) Suppose $x_{i+1} \geq x_i$. Then $(I-A)x_{i+1} \geq (I-A)x_i$ iff $x_{i+1} - x_i \geq Ax_{i+1} - Ax_i \geq 0$. Also $|(I-A)x_{i+1} - (I-A)x_i| = |Ax_i - Ax_{i+1}| \leq |x_{i+1} - x_i|$.

A similar argument completes the proof if $x_{i+1} \leq x_i$. □

Corollary. A is ftp iff $I - A$ is ftp.

What is important to note is that L_n and U_n are ftp but M_1 is not, as the example $x = \langle x_i = (-1)^i \rangle$ demonstrates, when $x_i \geq x_{i+1}$ implies $M_1 x_i < M_1 x_{i+1}$ and also $|M_1 x_i - M_1 x_{i+1}| = 2|x_i - x_{i+1}|$. Although the sequence is not in ℓp, the example is relevant as a finite part can be appended with zeroes to bring it into ℓp.

In the more general theory of Morphological Filters (29) there are several related filters that are useful, for practical and theoretical purposes. It may be useful to consider some of them here in the context of $LULU$-operators. Serra calls syntone operators increasing and defines a filter (Morphological-) as a syntone and idempotent operator. He calls F extensive if $I \leq F$ and anti-extensive if $G \leq I$. An extensive filter is called a closing and an anti-extensive filter an opening . An underfilter F has $F^2 \leq F$ and an overfilter $F^2 \geq F$. A v-underfilter F has $F(F \vee I) = F$ and a $\wedge$-filter has $F(F \wedge I) = F$, and strong filter both. The (Morphological-) center of two filters G and F with $F \geq G$ is defined as $B = (F \wedge I) \vee G = (F \vee G) \wedge (I \vee G) = F \wedge (I \vee G)$. (In these definitions $(A \vee B)x_i = (Ax_i) \vee (Bx_i) = \max\{Ax_i, Bx_i\}$ and $(A \wedge B)x_i = (Ax_i) \wedge (Bx_i) = \min\{Ax_i, Bx_i\}$.)

In particular L_n and U_n are strong filters and their morphological center is I, and L_nU_n and U_nL_n are strong filters and their center is idempotent. It is also co-idempotent, provided $(L_nU_n)\vee I$ and $(U_nL_n\wedge I)$ are co-idempotent or provided $(U_nL_n)\vee I$ and $(L_nU_n)\wedge I)$ are co-idempotent. See Wild (52) for a proof that the center is co-idempotent. The rest follows already from Serra.

Theorem 6.11. If A and B are ftp then $A\wedge B$ and $A\vee B$ are ftp and the morphological center, if it exists, is ftp.

Proof. (a) Let $W = A\vee B$ and $x_{i+1}\geq x_i$. Then $(A\vee B)x_{i+1}\geq Ax_{i+1}, Bx_{i+1}$. Since A and B are ftp, $(A\vee B)x_{i+1}\geq Ax_i, Bx_i$ and therefore $(A\vee B)x_{i+1}\geq (Ax_i)\vee(Bx_i) = (A\vee B)x_i$. But then also $0\leq (A\vee B)x_{i+1}-(A\vee B)x_i = Ax_{i+1}-(A\vee B)x_i$, if $Ax_{i+1}\geq Bx_{i+1} = Bx_{i+1}-(A\vee B)x_i$, if $Ax_{i+1}\leq Bx_{i+1}$ and $|(A\vee B)|x_{i+1}-(A\vee B)x_i|\leq Ax_{i+1}-Ax_i\leq x_{i+1}-x_i$ or $|(A\vee B)x_{i+1}-(A\vee B)x_i|\leq Bx_{i+1}-Bx_i\leq x_{i+1}-x_i$. A similar argument holds if $x_{i+1}\leq x_i$.

(b) Let $W = A\wedge B$ and $x_{i+1}\geq x_i$. Then $(A\wedge B)x_i\leq Ax_i, Bx_i$. But A and B are ftp, so that $Ax_i\leq Ax_{i+1}$ and $Bx_i\leq Bx_{i+1}$. Therefore $(A\wedge B)x_i\leq Ax_{i+1}\wedge Bx_{i+1} = (A\wedge B)x_{i+1}$ and also

$$\begin{aligned}0\leq (A\wedge B)x_{i+1}-(A\wedge B)x_i &= (A\wedge B)x_{i+1}-Ax_i,\ \text{if } Ax_i\leq Bx_i\\ &= (A\wedge B)x_{i+1}-Bx_i,\ \text{if } Bx_i\leq Ax_i^{\cdot}\end{aligned}$$

Therefore

$$\begin{aligned}|(A\wedge B)x_{i+1}-(A\wedge B)x_i| &\leq Ax_{i+1}Ax_i,\ \text{if } Ax_i\leq Bx_i\\ &\leq Bx_{i+1}-Bx_i,\ \text{if } Bx_i\leq Ax_i.\end{aligned}$$

Both quantities on the right are not larger than $x_{i+1}-x_i$. (A similar argument again holds if $x_{i+1}\leq x_i$.)

(c) The morphological center, if it exists, is of the type $A\vee(B\wedge I)$ or $A\wedge(B\vee I)$, both of which are compositions of ftp operators. □

Theorem 6.12. Let $x\in\mathcal{M}_n$ and A ntp. Then $Ax\in\mathcal{M}_n$.

Proof. Consider $x\in\mathcal{M}_n$. Then $\{x_i, x_{i+1},\dots,x_{i+n},x_{i+n+1}\}$ are monotone for each i. Thus $x_i\leq x_{i+1}\leq\cdots\leq x_{i+n+1}$ or $x_i\geq x_{i+1}\geq\cdots\geq x_{i+n+1}$. In both cases the output of A inherits the inequalities, so that $\{Ax_i, Ax_{i+1},\dots,Ax_{i+n+1}\}$ is monotone, for each i. This means that $Ax\in M_n$. □

Corollary. If A is ntp, then, for all $j\leq n$, since L_j and U_j preserve n-monotone sequences,

$$\begin{aligned}L_jAL_nU_n &= U_jAL_nU_n = AL_nU_n,\\ L_jAU_nL_n &= U_jU_nL_n = AU_nL_n,\\ M_jAU_nL_n &= M_jAL_nU_n = AL_nU_n.\end{aligned}$$

The corollary is very useful for simplifying compositions of $LULU$-operators and other ftp, operators, and can be partially generalized to equalities like

$L_nAL_n = AL_n$ etc, using the concepts of upward n-monotone and downward n-monotone (24). We consider the definition and basic result although the important corollary is easily proved directly.

Definition. For $k \geq 1$ call a segment $(x_i, x_{i+1}, \ldots, x_{i+k}, x_{i+k+1})$ of a series $x \in X$ a k-upwards arc (resp. k-downwards arc) if $x_i > x_{i+1} = \cdots = x_{i+k} < x_{i+k+1}$ (resp. $x_i < x_{i+1} = \cdots = x_{i+k} > x_{i+k+1}$. Let $n \geq 0$. We say that $x \in X$ is upwards n-monotone (downwards n-monotone) if $k \geq n+1$ for each k-upwards (k)-downwards) arc contained in x.

Theorem 6.13. Let A be fully trend preserving (ftp). Then A preserves upwards n-monotone and downwards n-monotone sequences.

Proof. The invariant sequences of U_n (respectively L_m) are exactly the upwards n-monotone (respectively downwards m-monotone) sequences in ℓ_p. Let x be upwards n-monotone, and $Ax_i > Ax_{i+1} = \cdots = Ax_{i+k} < A_{i+k+1}$ be an upward arc of A, and $k < n+1$. Then $x_i > x_{i+1}$ and $x_{i+k} < x_{i+k+1}$ (otherwise contradictions arise). Suppose x_j, with $i+1 < j < i+k+1$ is such that it differs from a neighbor. Then an arc with smaller support results. Repeating the argument yields an arc with $x_j > x_{i+1} = \cdots = x_{j+\ell} < x_{j+\ell+1}$ and $\ell < k < m+1$. This contradicts the fact that x is upward n-monotone. A similar argument for downward n-monotone sequences proves the theorem. □

Corollary. If A is ftp, then for all $j \leq n$, $L_jAL_n = AL_n$ and $U_jAU_n = AU_n$.

This corollary is the important result here, as in the analysis of smoothers, the following composition of the basic *LULU* smoothers L_n and U_n with different n, is often useful.

Definition. $C_n = L_nU_nC_{n-1}$ and $F_n = U_nL_nF_{n-1}$, with $C_0 = F_0 = I$.

For simplification it is convenient to note that if $n = \max\{k, m\}$, then $U_kU_m = U_n$ and $L_kL_m = L_n$. A similar result holds for C_n and F_n.

Theorem 6.14. $C_mC_k = C_n$ and $F_mF_k = F_n$, where $n = \max\{k, m\}$.

Proof. Suppose $m \leq k$. The output of C_k is k-monotone and C_m preserves this. Thus C_m cannot change the output and $C_mC_k = C_k$. Suppose $m = k+1$. $C_mC_k = C_{k+1}C_k = L_{k+1}U_{k+1}C_kC_k = L_{k+1}U_{k+1}C_k$, by the first part of the proof. But this is C_{k+1}, by definition. By induction it follows that the theorem is also true for any $m > n$. Clearly a similar proof holds for the other part of the theorem. □

The fact that the output of C_n is n-monotone implies also that the $F_nC_n = C_n$ and $C_nF_n = F_n$, since all the operators involved are syntone. It follows by induction that

$$U_nL_n \leq F_n \leq C_n \leq L_nU_n,$$

since $C_n = L_nU_nC_{n-1} \leq L_nU_nL_{n-1}U_{n-1} \leq L_nU_nU_{n-1} = L_nU_n$, by the swallowing theorem $U_nU_j = U_n$ for $j \leq n$.

In the following graph a section of a random sequence x together with images of ftp operators $L_1U_1, I - L_1U_1, U_1L_1, I - U_1L_1, C_2, (I - C_2)C_1, F_2, (I - F_2)F_1, C_3, (I - C_3)C_2, F_3, (I - F_3)F_2$ are plotted on the same graph, demonstrating the trend preservation. Lines are drawn to connect consecutive values of the sequences for visibility and circles identify x, which has the largest variation.

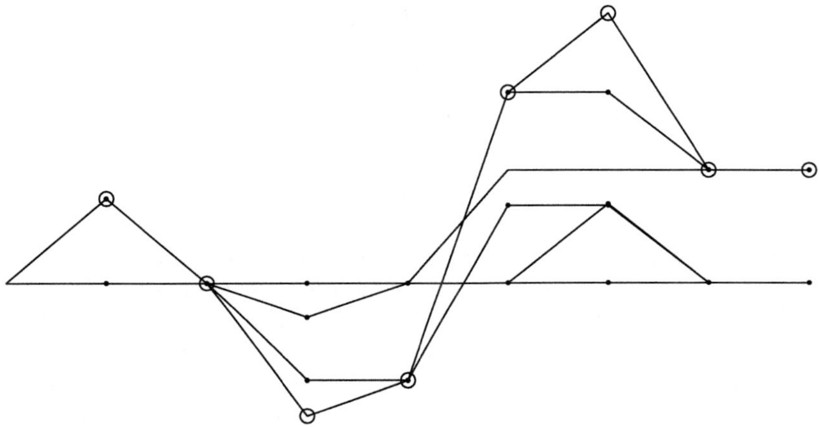

Figure 6.4. The output of x for several ftp operators to illustrate local trend preservation.

Theorem 6.15. Let $x \in \mathcal{M}_{n-1}$ and A ftp. Then

$$U_n(I - AU_n)x = U_nx - U_nAU_nx \quad \text{and} \quad L_n(I - AL_n)x = L_nx - L_nAL_nx.$$

Proof. Consider a sequence $x \in \mathcal{M}_{n-1}$ and index i. Assume first that $Ux_i = U_nx_i = x_i$. Since $Ux_i = \min\{\max\{x_{i-n}, \ldots, x_i\}, \ldots, \max\{x_i, x_{i+1}, \ldots, x_{i+n}\}\}$, there is an index $j \in [i-n, i]$ such that $\max\{x_j, \ldots, x_{j+n}\} = x_i$. This means that

$$x_j \leq x_{j+1} \leq \cdots \leq x_{i-1} \leq x_i \geq x_{i+1} \geq \cdots \geq x_{j+n},$$

since $x \in \mathcal{M}_{n-1}$ means that the consecutive values are monotone. The operators U, AU and $I - AU$ are ftp, so that the corresponding values of the outputs inherit this order. Therefore $U(I - AU)x_i = (I - AU)x_i = x_i - AUx_i = Ux_i - UAUx_i$.

Assume therefore that $Ux_i \neq x_i$, and therefore $Ux_i > x_i$. Since $x \in \mathcal{M}_{n-1}$ each of the sets $\{x_k, x_{k+1}, \ldots, x_{k+n}\}$ involved is monotone, and each has a maximum larger than x_i, the sequence must have a subset such that

$$x_j > x_{j+1} = x_{j+2} = \cdots = x_i = x_{i+1} = \cdots = x_{j+n} < x_{j+n+1}.$$

Because the relevant operators are ftp, the outputs of the operators U, AU and $I - AU$ must inherit the order structure, except for the strict inequalities weakening to inequalities. Noting that $Ux_i = \min\{x_j, x_{j+n+1}\}$ it is clear that similarly

$UAUx_i = \min\{AUx_j, AUx_{j+n+1}\}$ and $U(I-AU)x_i = \min\{(I-AU)x_j,(I-AU)x_{j+n+1}\}$.

Furthermore, since $\max\{x_j,\ldots,x_{j+n}\} = x_j$ and $\max\{x_{j+1},\ldots,x_{j+1+n}\} = x_{j+n+1}$, it follows that $Ux_j = x_j$, $Ux_{j+n+1} = x_{j+n+1}$, and similarly

$$\begin{aligned} UAUx_j &= AUx_j, & UAUx_{j+n+1} &= AUx_{j+n+1},\\ U(I-AU)x_j &= (I-AU)x_j, & U(I-AU)x_{j+n+1} &= (I-AU)x_{j+n+1}. \end{aligned}$$

Therefore

$$\begin{aligned} U(I-AU)x_j &= \min\{(I-AU)x_j,(I-AU)x_{j+n+1}\}\\ &= \min\{x_j - AUx_j, x_{j+n+1} - AUx_{j+n+1}\}\\ &= \min\{Ux_j - UAUx_j, Ux_{j+n+1} - UAUx_{j+n+1}\}. \end{aligned}$$

One of the values x_j, x_{j+n+1}, say x_j for convenience, is equal to $Ux_i = Ux_{i-1} = \cdots = Ux_j$. AUx inherits this sequence of equalities and so does $UAUx_j$. Therefore $U(I-AU)x_i = \min\{Ux_i - UAUx_i, Ux_{j+n+1} - UAUx_{j+n+1}\} = Ux_i - UAUx_i$.

The last step follows from the fact that $I - UA$ is ftp, so that $Ux_{j+n+1} - UAUx_{j+n+1} = Ux_i + (Ux_{j+n+1} - Ux_i - UAUx_{j+n+1} + UAUx_i) - UAUx_i \geq Ux_i - UAUx_i$. This is because the term in brackets is positive, since $Ux_i = Ux_{j+n}$ and $UAUx_i = UAUx_{j+n}$ and $Ux_{j+n+1} - Ux_{j+1} \geq UA(Ux)_{j+n+1} - UA(Ux)_{j+n}$, as UA is ftp.

Thus the theorem is proved for the first statement. The other case follows from a similar argument with L_n instead of U_n. □

Theorem 6.16. If A is ftp and $x \in \mathcal{M}_{n-1}$,

$$\begin{aligned} L_nU_n(I - AL_nU_n)x &= L_nU_nx - L_nU_nAL_nU_nx \quad \text{and}\\ U_nL_n(I - AU_nL_n)x &= U_nL_nx - U_nL_nAU_nL_nx. \end{aligned}$$

Proof. It follows from the previous theorem that

$$\begin{aligned} L_nU_n(I - AL_nU_n)x &= L_n(U_nx - U_nAL_nU_nx) = L_n(I - U_nAL_n)U_nx\\ &= (L_n - L_nU_nAL_n)U_nx = L_nU_nx - L_nU_nAL_nU_nx. \end{aligned}$$ □

Theorem 6.17.

$$\begin{aligned} C_j(I - C_n) &= C_j - C_n, \quad \text{for } j \leq n \text{ and}\\ F_j(I - F_n) &= F_j - F_n, \quad \text{for } j \leq n. \end{aligned}$$

Proof. Let $j = 1$. Then $C_1 = L_1U_1$ and $C_n = C_nL_1U_1$ so that,

$$\begin{aligned} C_1(I - C_n) &= L_1(U_1(I - C_nL_1U_1))\\ &= L_1(U_1 - U_1C_nL_1U_1), && \text{by Theorem 1, as } C_nL_1 \text{ is ftp}\\ &= L_1(I - U_1C_nL_1)U_1\\ &= (L_1 - L_1U_1C_nL_1)U_1, && \text{by Theorem 1, as } U_1C_n \text{ is ftp}\\ &= L_1U_1 - C_n = C_1 - C_n. \end{aligned}$$

The theorem is therefore true for $j = 1$. The rest of the proof follows by induction. Assume it is true for $j = m < n$. Then

$$\begin{aligned} C_{m+1}(I - C_n) &= L_{m+1}U_{m+1}C_m(I - C_n) = L_{m+1}U_{n+1}(C_m - C_n) \\ &= L_{m+1}U_{m+1}(C_m - C_nC_m) = L_{m+1}U_{m+1}(I - C_n)C_m \\ &= (L_{m+1}U_{m+1} - L_{n+1}U_{m+1}C_n)C_m \\ &= L_{m+1}U_{m+1}C_m - L_{n+1}U_{m+1} + C_n \\ &= C_{m+1} - C_n, \quad \text{if } m+1 \leq n. \end{aligned}$$

A similar proof holds for the second part with L and U interchanged. □

Corollary. C_n and F_n are co-idempotent.

The co-idempotence of more general compositions of different L_iU_i operators may also hold, but this is not the concern here.

Example 1: $L_3U_3L_1U_1(I - L_3U_3L_1U_1) = L_3U_3(I - L_1U_1L_3U_3)L_1U_1$.

Whether this simplifies to $L_3U_3(I - L_3U_3)L_1U_1 = 0L_1U_1 = 0$, is as yet unclear.

Theorem 6.18. The operators $C_n, C_m, I - C_n$ and $I - C_m$ all commute for all positive integers n and m.

Proof. (i) $C_nC_m = C_{\max\{n,m\}}$, by Theorem 4.

(ii) $C_n(I - C_m) = C_n - C_m$, for all $n \leq m = C_n - C_mC_n = (I - C_m)C_n$, $C_n(I - C_m) = C_nC_m(I - C_m)$, for all $n \geq m = C_nO$, by the co-idempotence of C_m. Also $(I - C_m)Cn = C_n - C_mC_n = C_n - C_n = 0$.

(iii) $(I - C_n)(I - C_m) = I - C_m - C_n(I - C_m) = I - Cm - (I - C_m)C_n = I - C_m - C_n + C_{\max\{n,m\}} = I - C_{\min\{n,m\}}$ □

Corollary. The operators F_n and $I - F_n$ all commute. (The proof can be done as before, or using the duality $C_n(-x) = -F_n(x)$.)

An operator A that is idempotent and co-idempotent is a separator in that it consistently separates (decomposes) a sequence x into Ax and $(I - A)x$. This is as near to a projection as we can hope to get a nonlinear operator. Since even the composition of two idempotent smoothers does not have to be idempotent, it is of crucial importance that smoothers that are selected to smooth sequentially are as consistent as possible. Since U_n and L_n are the primary smoothers under consideration here, the idempotence and co-idempotence of the basic compositions like L_nU_n, U_nL_n, C_n and F_n go a long way towards consistent behavior. For further consistency the following results are also important.

Theorem 6.19. Let $x \in \mathcal{M}_{n-1}$, and A, B ntp. For all $\alpha, \beta \geq 0$:

(a) $U_n(\alpha A + \beta BU_n)x \geq \alpha U_n Ax + \beta BU_n x$ and $L_n(\alpha A + \beta BL_n)x \leq \alpha L_n Ax + \beta BL_n x$.
(b) If A also commutes with U_n on such a sequence, then $U_n(\alpha A + \beta BU_n)x = \alpha U_n Ax + \beta BU_n x$.
(c) If A also commutes with L_n on such a sequence, then $L_n(\alpha A + \beta BL_n)x = \alpha L_n Ax + \beta BL_n x$.

Proof. (a) With the notation $U = U_n$, let $x \in \mathcal{M}_{n-1}$ and i an index. Assume first that $Ux_i = x_i$. Then there is an index $j \in [i-n, i]$ such that $\max\{x_j, \dots, x_{j+n}\} \leq x_i$. But the sets $\{x_j, \dots, x_i\}$ and $\{x_i, x_{j+n}\}$ are monotone, thus

$$x_j \leq x_{j+1} \leq \cdots \leq x_i \geq \cdot \geq x_{j+n}.$$

$A, BU, UA, UBU, \alpha A + \beta B$ and $U(\alpha A + \beta BU)$ all transfer these inequalities, to their output, since they are ntp. Therefore

$$UAx_i = Ax_i, UBUx_i = BUx_i \quad \text{and}$$

$U(\alpha A + \beta BU)x_i = \alpha Ax_i + \beta BUx_i = \alpha UAx_i + \beta UBUx_i$. Assume therefore that $U_n x_i \neq x_i$ (and therefore $U_n x_i > x_i$). As previously argued there is a $j \in [i-n, i]$ such that

$$x_{j-1} > x_j = x_{j+1} = \cdots = x_i = \cdots = x_{j+n-1} < x_{j+n},$$

and

$$\begin{aligned} Ux_{j-1} \geq Ux_i &= \min\{x_{j-1}, x_{j+n}\} \\ &= Ux_{j-1} = \cdots = Ux_i = \cdots = Ux_{j+n-1} \leq Ux_{j+n}. \end{aligned}$$

All the operators in question are ntp so that again the inequalities are inherited by the outputs and

$$\begin{aligned} UAx_i &= \min\{Ax_{j-1}, Ax_{j+n}\}, \\ UBUx_i &= \min\{BUx_{j-1}, BUx_{j+n}\} = BUx_i, \end{aligned}$$

since B is ntp.

$$\begin{aligned} U(\alpha A + \beta BU)x_i &= \min\{(\alpha A + \beta BU)x_{j-1}, (\alpha A + \beta BU)x_{j+n}\} \\ &= \min\{\alpha Ax_{j-1} + \beta BUx_{j-1}, \alpha Ax_{j+n} + \beta BUx_{j+n}\}. \end{aligned}$$

Now $\alpha UAx_i + \beta BUx_i \leq \alpha Ax_{j-1} + \beta BUx_{j-1}$ and $\alpha UAx_i + \beta BUx_i \leq \alpha Ax_{j+n} + \beta BUx_{j+n}$. The inequality therefore also holds for the minimum of the two, proving that $\alpha UAx_i + \beta BUx_i \leq U(\alpha A + \beta B)x_i$.

A similar argument proves the other part of (a).

(b) Noting that $UAx_i = \min\{Ax_{j-1}, Ax_{j+n}\}$, assume that $Ax_{j-1} \leq Ax_{j+n}$ so that $UAx_i = Ax_{j-1}$. (A similar argument holds if the other is smaller.) Then $Ux_{j-1} = Ux_j = \cdots = Ux_i = \cdots = Ux_{j+n-1} \leq Ux_{j+n}$ Since A and B are ntp the equalities are inherited by AUx and BUx so that $AUx_i = AUx_{j-1} \leq AUx_{j+n}$ and $BUx_i = BUx_{j-1} \leq BUx_{j+n}$, and so that

$$\begin{aligned} U(\alpha A + \beta BU)x_i &\leq \alpha Ax_{j-1} + \beta BUx_{j-1} \\ &\leq \alpha UAx_{j-1} + \beta BUx_i. \end{aligned}$$

If now $UAx_{j-1} = AUx_{j-1}$ then since $AUx_{j-1} = AUx_i$, we get $U(\alpha A + \beta BU)x_i \leq \alpha AUx_i + \beta BUx_i = \alpha UAx_i + \beta BUx_i$. This, together with the inequality of part (a) of the proof yields the result $U(\alpha B + \beta BU)x_i = \alpha UAx_i + \beta BUx_i$, which proves part (b) of the theorem.

(c) A similar proof to the above, or using a duality argument, yields the required equality. □

Theorem 6.20. For $x \in \mathcal{M}_{n-1}$ and $\alpha, \beta \geq 0$:

$$\begin{aligned} U_n(\alpha(I - L_nU_n) + \beta L_nU_n)x &= \alpha U_n(I - L_nU_n)x + \beta L_nU_nx \quad \text{and} \\ L_n(\alpha(I - U_nL_n) + \beta U_nL_n)x &= \alpha L_n(I - U_nL_n)x + \beta U_nL_nx, \\ L_nU_n(\alpha(I - L_nU_n) + \beta L_nU_n)x &= \beta L_nU_nx \quad \text{and} \\ U_nL_n(\alpha(I - U_nL_n) + \beta U_nL_n)x &= \beta U_nL_nx. \end{aligned}$$

Proof. L_n and $I - L_nU_n$ are ftp and thus ntp. By Theorem 6.15 $I - L_nU_n$ commutes with U_n, and by Theorem 6.19(b) the first equality holds. A similar proof holds for the second equality.

For the third equality we apply the first equality. Then

$$\begin{aligned} L_nU_n(\alpha(I - L_nU_n) + \beta L_n(U_n)x &= L_n[\alpha U_n(I - L_nU_n)x + \beta L_nU_nx] \\ &= L_n[\alpha(U_nx - U_nL_nU_nx) + \beta L_nU_nx], \quad \text{by Theorem 6.15.} \\ &= L_n[\alpha(I - L_n) + \beta L_n]U_nx, \quad \text{since } U_nL_nU_n = L_nU_n. \end{aligned}$$

Since L_n commutes with $I - L_n$, we have by Theorem 6.19 that the last expression becomes

$$[\alpha L_n(I - L_n) + \beta L_n]U_nx = \beta L_nU_nx,$$

since L_n is co-idempotent.

The last equality is proved similarly, or by duality. □

The theorems above that have resulted from the simple concept of ntp turn out to be extremely strong. Recalling the problem of selecting good smoothers amongst the n-LULU similar ones (those that lie inside the LULU interval $[U_nL_n, L_nU_n]$), the natural candidates now seem to identify themselves, if our understanding of the behavior is to be optimal. L_nU_n and U_nL_n themselves had all the nice properties, but in practice it becomes clear that, for large n, their bias is excessive. (They are too far from the center of the $LULU$-interval). Furthermore it is often natural to smooth successively, in order to decide how much smoothing is required. Philosophically, we know that smoothing destroys information, unless that which is removed is stored separately. A smoother is designed to selectively remove unwanted information to highlight wanted information. This is why it is considered an art. If some form of automated selection of degree of smoothing of data is required, it seems natural to smooth recursively, until some selected criterion is met. This leads naturally to the choice of F_n or C_n, for some n, (to be selected as the computation progresses). We shall find that computationally this does not become excessive. Such a recursive removal of successive levels of "noise" in a systematic way immediately brings to mind a similar procedure; that of Multiresolution Analysis in Wavelet Theory. This idea is worth investigating in the context of $LULU$-theory.

7. Multiresolution Analysis of Sequences

"Treu der Natur!" – wie fängt er's an;
Wann wäre je Natur im Bilde abgetan
Unendlich ist das kleinste Stück der Welt! –
Er malt zuletzt davon, was ihm gefällt
Und was gefällt ihm? was er malen kann.

Nietzsche

The Fast Fourier Transform (FFT) has often been called the most important mathematical tool in modern technology. In a similar way the Fast Wavelet Transform (FWT) may have an impact, for instance, in image processing and transmission. A different idea has been pursued, and is considered to be even better in the "Multiresolution Analysis" of digital images. In the book, Image Processing and Data Analysis, Stark Murtagh and Bijaoui follow a discussion of the Wavelet Transform with a section on Multiresolution Analysis (MRA) based on the Median Transform. Roughly speaking, the averaging filter that maps a function onto a function in a space of lower dimension in the Wavelet Transform is replaced by a median smoother. As usual this can heuristically be motivated by yielding a more robust estimator of an average and therefore outliers do less damage on the "smoother component" in the mapping. Iterative and non-iterative algorithms for Median Transforms are presented.

The claim is made that this MRA is well suited when image reconstruction is done from a subset of the (additive-) decomposition for purposes of restoration, compression and partial reconstruction. The reconstructed image is often found to have fewer artifacts than in the case of wavelet decomposition. These artifacts are often in the form of the specific wavelets chosen. An example is the negative ring surrounding bright point sources. Shapes are found to be closer to those of the input image. The claim is that this is due to the nonlinearity of the median filter (strictly speaking the nonlinearity permits this better preservation). Computational requirements are listed as high, although there is a saving in that these transforms can be performed in integer values only and decimation can yield a considerable economization. Other morphological tools, specifically N erosions followed by N dilations are mentioned, with the observation that results were found to be better with the median. Morphological transforms are presented, with mention of the good estimate of the image background that is obtained, especially for images with small structures, as in astronomy.

In one-dimensional signal analysis the origin of many artifacts (irritating significant distortions) in signals partially reconstructed from wavelet decompositions, can often be understood in the context of approximation theory. The linear projections used in wavelet decompositions may perform relatively well in audio signals. The explanation may ultimately be physiological, but for the purpose at hand, it is sufficient to observe that an audio signal is often well approximated locally by trigonometric polynomials, so that the FFT can be truncated for compression. Such local approximation does some damage (Gibbs' phenomenon), but "softer" windows can be employed to lessen this. Depending upon the support of a specific wavelet, and the norm of an associated mapping onto a subset of lesser dimension, the Gibbs' phenomenon may be acceptable for the partial reconstruction.

Visual signals or images have a different problem, also in one dimension. Sharp edges and constant regions play a significant part in the acceptability or recognisability of an image. Such data, approximated by smooth functions, or sequences of sampled smooth functions, bring Gibbs into play, with possible "overshoot" and "undershoot" near edges. An edge can be interpreted as having an impulse in the derivative, or an impulse in the sequence of differences of samples of such an image. Impulses are not handled well by linear filters, or smooth functions in approximation theory. Convoluting a wavelet with an impulse yields precisely the wavelet concerned so that impulsive noise on a sequence can be expected to yield spurious features, often close to the shape of the wavelets, or sums of these, at all resolution levels. In the extreme case of the non-local Fourier Transform, the impulse results in a constant function in the frequency domain. This contamination is difficult to remove afterwards. A linear filter preserves the "energy" in an impulse. Even in the case of the simplest Haar-wavelet decomposition, where the projection onto the lower frequency sequence has minimal norm, an impulse is merely spread onto all levels of the wavelet decomposition.

With the $LULU$-theory an alternative arises. There are operators with pulses as eigensequences, and where signals are naturally composed of pulses of various widths, a corresponding pulse-decomposition/trend-decomposition can be envisaged. In modern control theory, digital transmission theory, chromatography, image processing, to name but a few examples, it is often natural to assume a composition of pulses inherent in a measurement. In earthquakes there can be a natural assumption of pulses of vibrations, which can be handled by wavelets, pulse decomposition or a combination. In such cases the pulse can also be considered to be in the frequency domain.

Due to the inherent ambiguity in the concept of pulses, the decomposition has to be nonlinear, and the natural framework of analysis and computation may by the $LULU$-theory. The nonlinearity is not to be seen merely as a generalization of the linear case, but a fundamentally different perspective, with a very strong underlying logical structure. This fundamentally specific use of the term "nonlinear" is often insufficiently appreciated by precisely those that are very familiar with linear signal processing.

The idea of Multiresolution Analysis with pulses, or Trend Analysis is best developed by considering the simplest Spline Wavelet Analysis, or Haar-decomposition, as a comparison. For a background any textbook on wavelet theory can be consulted, for instance (8).

For the purpose of introducing and motivating the subsequent results we exploit the simple one-to-one correspondence of a sequence as a zero-degree B-spline function sampled at the centers of intervals. Local monotonicity as a concept of smoothness then automatically comes into play in both the wavelet decomposition and the $LULU$-decompositions.

Given a function f, a subset of a function space is chosen so that it is spanned by translations of a so-called scaling function φ, which is itself a linear combination of the type $\phi(t) = \sum_{i=-\infty}^{\infty} \alpha_i \phi(2t-i)$. Generalizations are possible but not required here, in fact it will be sufficient for the purpose at hand to consider only the simple Haar-wavelet decomposition. The "scaling function" ϕ in this case is simply the characteristic function of the interval $[0,1)$, thus a B spline of order 1 (degree 0). It is clear that $\phi(t)$ is a linear combination of $\phi(2t)$ and $\phi(2t-1)$, and that in this case it is simply the sum. Thus $\phi(t) = \phi(2t) + \phi(2t-1)$, and by induction $\phi(t) = \sum_{i=0}^{2^k} \phi(2^k t - i)$.

Consider the space, which is the span of these functions, and consider a sequence $x = \{x_i : i = 0, \dots, N-1\}$ which is a sampling of a function of f at the values $t_i = \dfrac{i+\delta}{2^k}$, where $0 \le \delta < 1$. Clearly, in this case, the sequence can be identified with the sequence of coefficients w.r.t. the basis $\{\phi_i; \phi_i(k) = \phi(2^k t - i)\}$. Letting B_j be the space such that

$$B_j = \text{ span}\,\{\phi_i; \phi_i(t) = \phi(2^j t - i)\},$$

a best least squares estimate Px from B_{k-1} to a function $x = \sum_{i=0}^{N-1} \alpha_i \phi(2^k t - i)$ from B_k is easily obtained as

$$Px(t) = \sum_{j=0}^{\frac{N}{2}} \frac{1}{2}(\alpha_{2j} + \alpha_{2j+1})\phi(2^{k-1}t - j)$$

and

$$(x - Px)(t) = \sum_{j=0}^{\frac{N}{2}} \frac{1}{2}(\alpha_{2j+1} - \alpha_{2j})\psi(2^{k-1}t - j),$$

where $\psi(t) = \varphi(2t-1) - \varphi(2t)$, is the Haar-wavelet.

The basis is orthonormal w.r.t. the inner product $(x, y) = \int_{-\infty}^{\infty} x(t)y(t)dt$, and the set of wavelets ψ_i span the orthogonal complement of B_{k-1}, if $\psi_i(t) = \psi(2^{k-1}t - i)$. For later comparison it is sufficient to note that there is a preservation of "energy" in the sequences in that

$$||\alpha||_2^2 = \sum_{i=-\infty}^{\infty} \alpha_i^2 = \sum_{i=-\infty}^{\infty} 2\left(\frac{\alpha_{2j} + \alpha_{2j+1}}{2}\right)^2 + \sum_{i=-\infty}^{\infty} 2\left(\frac{\alpha_{2j+1} - \alpha_{2j}}{2}\right)^2.$$

In Figure 7.1 the "scaling sequence" and "wavelet sequence" are depicted, and in Figure 7.2 the first stage of a decomposition.

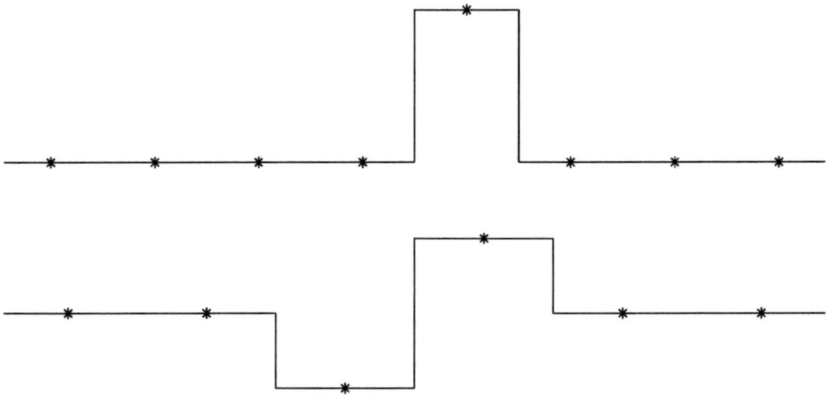

Figure 7.1. The Haar scaling functions and its wavelet of minimum support, identified with sequences.

As an example we can consider a digitized profile of a fort and its first decomposition.

Thus the sequence x is decomposed into a "smoother" sequence Px, which is pairwise constant, and a "rougher" sequence $x - Px$ which has pairwise elements equal in absolute value, but differing in signs. Significant, for later argument, is that Px has every three consecutive elements monotone.

Since the wavelet ψ has a definable frequency, it is natural to view the decomposition as a "smoother" spline Px and a wavelet component, which is a sequence with an associated frequency locally. Repeating such a decomposition, the original sequence x can eventually be decomposed into a constant sequence and several "layers" of wavelets at frequencies that are successively an octave lower than the previous. Schematically the wavelet decomposition can be viewed in the following diagram.

From the theory of wavelets the projections P_i have the following properties:

(i) $P_i f \perp \{f, P_1 f, \ldots, P_{i-1} f\}$,

(ii) P_i is idempotent, co-idempotent, linear and eigenvalues are only 0 and 1.

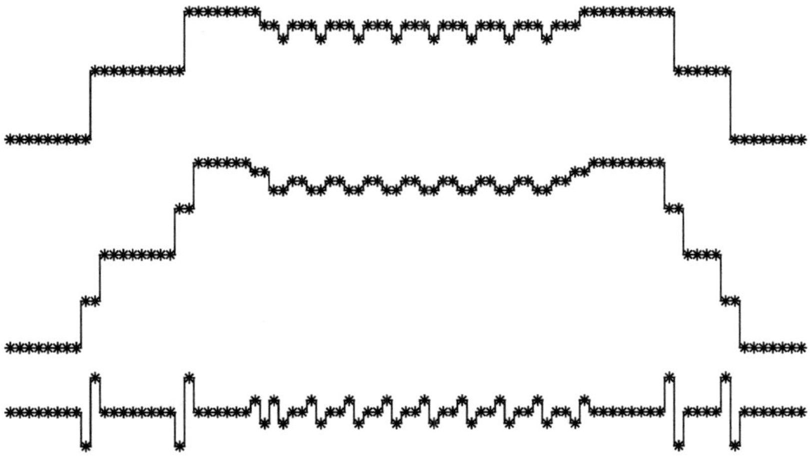

Figure 7.2. A sequence x decomposed into Px and $(I - P)x$.

Reconstruction can be achieved by adding the different "layers", and partial reconstruction by a subset of the layers, or a set of subsets of each layer. For data compression, for instance, the coefficients at each wavelet level can be "quantized" and only the nonzero quantized values stored or transmitted. Thus it is reasonable to speak of a "local frequency content" by considering the size of coefficients of wavelets with support at a chosen location.

What seems clear is that a signal with sections of almost constant value will have small frequency content at all frequencies there. Only in regions where the sequence varies significantly will there be a need for high frequency information being transmitted. For purposes of automatic analysis, significant changes in the sequences are identified with "wavelet activity" nearby. The effect seems local due to the small support of the scaling function φ and the wavelet ψ. In a sense therefore it is reasonable to say that global shape is determined by the low frequency content and higher resolution features reside in the wavelet coefficients. Hence the name Multiresolution Analysis (MRA). But should this name not be reserved for a stricter interpretation?

A linear transform can be characterized by its response to an "impulse". Letting d_i be the sequence $\{\delta_{ij} : j = 0, \ldots, N-1\}$, where δ_{ij} is the Kronecker-delta it is easy to see that, in spite of its minimal support, the wavelet decomposition will have exponential decaying amplitudes in exponentially growing support intervals. A partial reconstruction will therefore inevitably have deviation from the original sequence in an arbitrarily large region. This is the essential problem of the response of a linear mapping to "impulsive noise", if impulsive noise is precisely defined as an arbitrary multiple of a Kronecker-delta sequence. The "energy" in such an impulse is "spread" or "smeared", and, depending on the amplitude of the impulse, can completely swamp the essential signal in an arbitrarily large region.

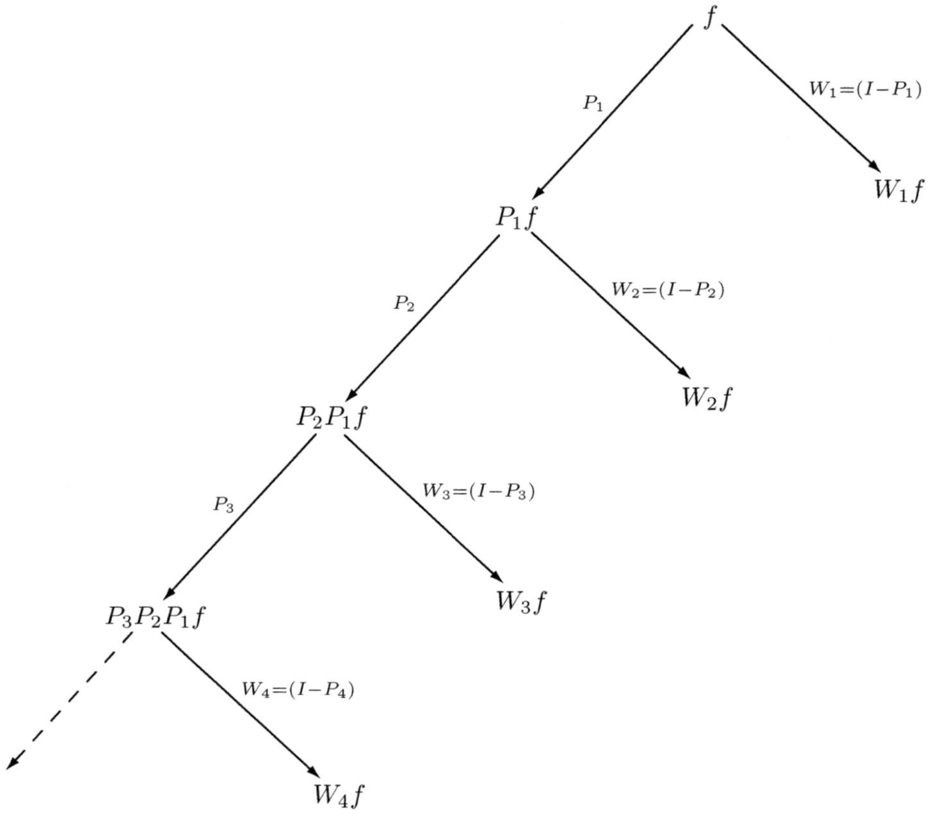

Figure 7.3. A schematic decomposition of a function f into wavelet sums.

If impulses like these can be expected in a measuring (or transmitting-) device, there will have to be some precaution taken, preferably before any linear transformation is performed. This led to the widespread use of pre-smoothing with running medians. The problem is essentially similar in all wavelet decompositions, and a progressively more local feature has progressively wider frequency content in general; the time-frequency window has an area exceeding a fixed positive quantity. Moreover, the damage done is not restricted to impulses, but to impulses in differences of the sequences too. Thus a sampling of a simple step function will generally have significant distortions spreading into all frequency levels. This behavior is also phase dependent.

It is illuminating to consider the above observations in the simple case of a sequence of samplings of a quadratic polynomial, with a simple unit step function (Heaviside function), uniformly distributed random noise and two isolated impulses (Kronecker-delta sequences) added. Choosing a Haar wavelet decomposition, the two successive least squares approximations from subspaces of a half and a quarter of the original dimension are depicted in Figure 7.4. In Figure 7.5 the "wavelet"-components of the first two frequency levels illustrate the geometric decay of the amplitude of features introduced by the impulses and the step discontinuity. In both cases two choices of spline knots are compared to show the phase dependence.

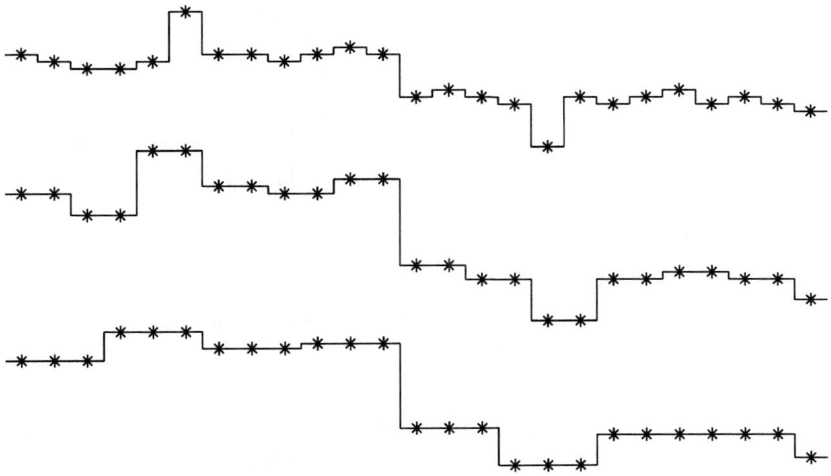

Figure 7.4. The original sequence and the sequences representing the partially reconstructed original sequence minus the highest frequency level and minus the two highest frequency levels.

In the above decompositions, it is clear that the synthetic wavelet activity due to the edge and the impulses is amplified by the factor α if the original impulses and jump discontinuity are, resulting in arbitrary synthetic features arbitrarily far in the decomposition.

In image processes, edges are significant for picture quality. Linear mappings generally do not preserve monotonicity in a sequence as Median smoothers do, and do so in a precise local sense, as do all rank order selectors. Can this be exploited? Clearly experience suggests that Multiresolution Analysis with medians works well in image processing (1). Are there computationally efficient alternatives, and can an underlying theory provide reassurance of predictable, comparable performance? And if a "feature" or a deviation from a "smooth" surrounding trend is at a precisely defined "resolution", in that it is sufficiently local, can it be separated without excessive contamination?

The concept of local monotonicity is a more compact characterization of the so-called "roots" of median smoothers. The behavior of M_n itself is enigmatic, and this can be associated with the existence of sections of "spurious" roots in a sequence. The "spurious" roots have recently been shown to be precisely the periodic ones (33) that are not in ℓ_p. In the case of M_2 it is essentially only one sequence, namely $x_i = (-1)^i$, and its multiples.

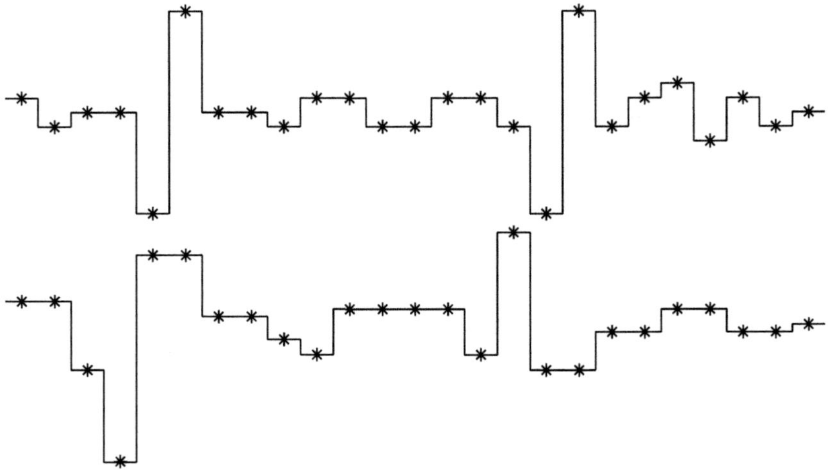

Figure 7.5. The sequences representing the highest and second highest frequency levels in the Haar-decomposition.

If $x \in X$ then x is 0-monotone and therefore $\mathcal{M}_0 = X$. Furthermore $\mathcal{M}_0 \supset \mathcal{M}_1 \supset \cdots \supset \mathcal{M}_n \supset \cdots$ form a sequence of nested subsets. (A Haar-decomposition projects into $\mathcal{M}_1$ with the first decomposition onto $\mathcal{M}_3$ with the second and so forth, and this is dependent on the phase, or therefore on the choice of the nodes of the splines involved.) If an alternative decomposition is to be constructed it can be considered prudent to aim for an elementary separator P in such a way that the criteria for a separator listed in Chapter 1 are not compromised too much.

Effectiveness: The output P must be a sequence without higher resolution detail.

Efficiency: The computations must be economical in terms of basic digital operations like logical comparisons, additions, multiplications, divisions etc.

Consistency: Mapping the output again should preserve it, or confirm it as good.

Stability: Input perturbations should not result in excessive output perturbations.

Considering the simple Haar-wavelet, the projection operator P, is effective since it is a projection onto a spline-subspace of half the original dimension of the (order 1 -) spline space at the sampling resolution. It is achieved efficiently by a simple averaging filter on the sequence. Since a projection operator is idempotent it preserves its own output. (It is noteworthy that this necessitates a filter that is not translation invariant (and therefore phase dependent as it depends on the nodes of the spline subspace of lower dimension onto which it is mapped.) Thus it does not meet the requirements of one of the axioms of a smoother, as introduced by Mallows. Stability has been argued to be suspect in the case of features that are "brief" impulses. Excessive output perturbations can result.

Choosing as the first separator in an alternative Multiresolution Analysis the operator $P_1 = L_1U_1$, it is clear that the output is 1-monotone. (U_1L_1 would be another choice leading to a similar scheme.) A general sequence $x \in X$ is then effectively mapped onto a 1-monotone sequence P_1x in $\mathcal{M}_1$. This operator is efficient, as will be shown later. Since L_1U_1 is idempotent it is consistent in preserving its output, but, since the operator is nonlinear, consistency demands somewhat more. Since P_1 is not a projection, the component $(I - P_2)x$ that is removed must also be consistently removed, in that $(I-P_1)(I-P_1)x = (I-P_1)x$.

This means that $I - P_1$ must also be idempotent, and this "co-idempotence" of $P_n = L_nU_n$ is equivalent to having $(I - P_n)x$ being a null-sequence of the operator P_n, for each x. The separation can thus be considered to be consistent. Stability is good since the operator has a Lipschitz constant, so that small amplitude perturbations cannot be amplified. Furthermore a single (large) impulse has an influence restricted to amplitudes of neighbors, and the influence is local. Under the heading of effectiveness, a further consideration arises when P is not a projection. A projection onto a subspace S is automatically a good approximation in the appropriate norm, and since the Lebesgue inequality is applicable, also if the norm of P is finite, in any other norm. If P_n is merely a separator, it is important to consider whether the image P_nx is a good approximation from $\mathcal{M}_n$ to the sequence x. This was shown to be so in the cases of L_nU_n and U_nL_n.

Thus the separator L_1U_1 effectively separates a sequence x into a good approximation L_1U_1x in $\mathcal{M}_1$ and a (high resolution) sequence $(I - L_1U_1)x$, which is a null-sequence of L_1U_1, and thus can be considered to consist of sufficiently local impulses, sufficiently separated not to yield a lower resolution nonzero output when mapped by L_1U_1. It can be considered as a sum of "noiselets". It must be stressed that the sum of such null-sequences is not automatically a null-sequence again! Furthermore, the operator U_nL_n, although considering L_nU_nx as a "signal", since it is in $\mathcal{M}_n$, does not necessarily consider (confirm) $(I - L_nU_n)x$ as being noise. This is because $U_nL_n(I - L_nU_n)$ is not the zero operator, although $L_nU_n(I - L_nU_n)$ is. L_nU_n and U_nL_n have a common range but $I - L_nU_n$ and $I - U_nL_n$ do not. These observations are associated with a fundamental "uncertainty principle" with respect to the concept of impulse, and will result in a similar uncertainty in the concept of resolution, if made strict. The two decompositions, with LU and UL can be effectively done in parallel, but separately.

In the following example it is instructive to view the two large impulses, as well as the "jump-discontinuity" as being sufficiently large, so that the "smoother" parts L_1U_1x and U_1L_1x will not be affected at all if the impulses are multiplied by an arbitrarily larger number. All the extra amplitude will be restricted to the noise-components $x - L_1U_1x$ and $x - U_1L_1x$. No change will result in lower layers of decompositions. Only when the amplitude decreases to the level of the local variation of the uncontaminated signal will there be a minor change in L_1U_1x and U_1L_1x, minor meaning of the order of the difference between the two. Since the Haar-decomposition is based on linear operators the multiplication of the impulses by an arbitrary α will result in a proportional distortion by the same factor in the wavelet component resulting in proportional distortion with exponentially increasing width. With sufficiently large amplitude impulses this can swamp all significant features of the original sequence.

In Figure 7.6 a sequence like that in Figure 7.4 is decomposed once by each of L_1U_1 and U_1L_1 to illustrate the claims made. This can conveniently be called a "trend decomposition". The Median transform would be between those of L_nU_1 and U_1L_1 and very similar.

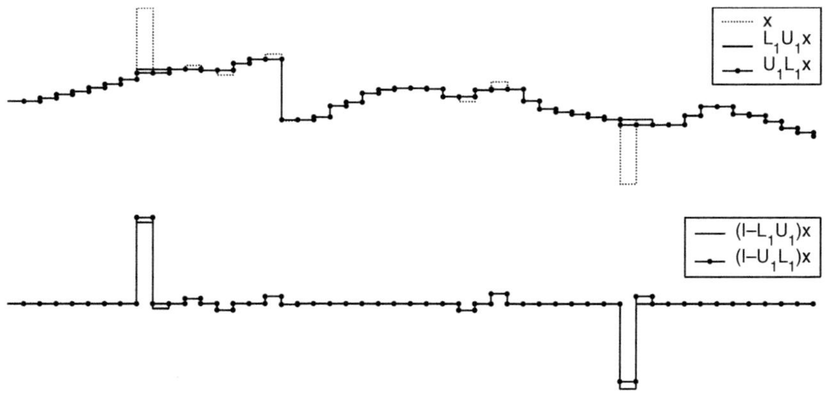

Figure 7.6. Illustration of a "trend decomposition" using the operators L_1U_1 and U_1L_1.

It is worthwhile comparing a Haar decomposition of the same sequence x.

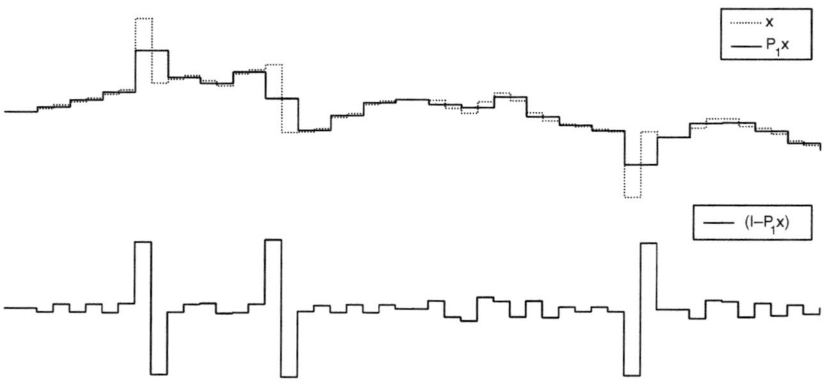

Figure 7.7. Haar (wavelet-) decomposition of the same sequence as above.

In the Haar-decomposition the large scale level change smears this significant feature to exponentially growing large sections of the successively smoother decompositions. The synthetic wavelet activity at the jump will have a correspondingly large amplitude in all subsequent wavelet layers. The well-known idea of thresholding the wavelet coefficient sequences to handle impulsive noise is clearly limited in its effectiveness, quite apart from the difficulty of choosing an appropriate threshold. The observations above are demonstrated with optimal clarity in the simple example of a constant signal and a single impulse of width 1. The first two Haar-decompositions in Figure 7.8 clearly demonstrate the exponentially growing width in both components of the decompositions, and because the operators involved are linear this behavior is scale independent. Omitting one or more levels of wavelet components will result in a large distortion in the reconstructed (-smoothed signal).

The $LULU$-decomposition of the same signal and impulse has the full energy of the impulse in its first noise-component and omitting of this (highest resolution) component and any other will leave a perfectly smoothed original constant signal. No sketch is required to see this.

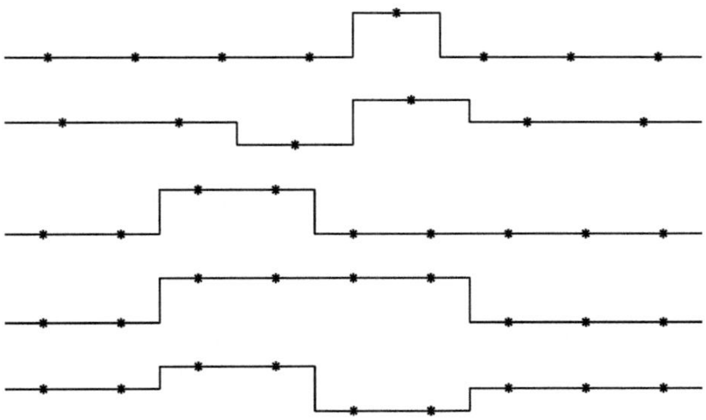

Figure 7.8. The impulse response of a Haar-decomposition.

Clearly, since the Haar decomposition is linear, this behavior will result in a similar distortion when added to any signal. Since the $LULU$-decompositions are not linear, the superposition of this impulse on a given signal x will not necessarily result in an undistorted removal of the impulse from x. But all the distortions will not exceed the magnitude of the local variation of the signal at the position of the impulse. Strictly speaking L_1U_1x and U_1L_1x cannot be distorted by more than the factor $\max\{|x_{i+1} - x_i|, |x_i - x_{i-1}|\}$ at i, if the arbitrarily large impulse is added at i, since both have either the value x_{i-1} or x_{i+1}. No distortion larger than this can result in any lower level of decomposition, since the induced 1-norms of all the operators L_nU_n and U_nL_n are 1. A further distortion occurs at the two neighboring points $i-1$ and $i+1$, but this distortion cannot exceed $|x_{i-2} - x_{i-1}|$ and $|x_{i+2} - x_{i+1}|$ respectively. Clearly the contamination can spread, but this cannot exceed the maximum amplitude of the signal in the corresponding region. Furthermore, it does not necessarily spread as far as it can theoretically, depending upon the given underlying signal. A precise analysis is not intended here, and may become exceedingly difficult. Experience suggests very limited growth in the contamination support.

A practical typical comparison that is illustrative of the specific advantage of the $LULU$-decomposition argued above is the following example, where a broad n-impulse is decomposed by $LULU$-decomposition and Haar decomposition. Except for the, progressively more unlikely event, as n increases, of the n-pulse starting and ending at a node of the lower dimensional spline subspaces, this is typical.

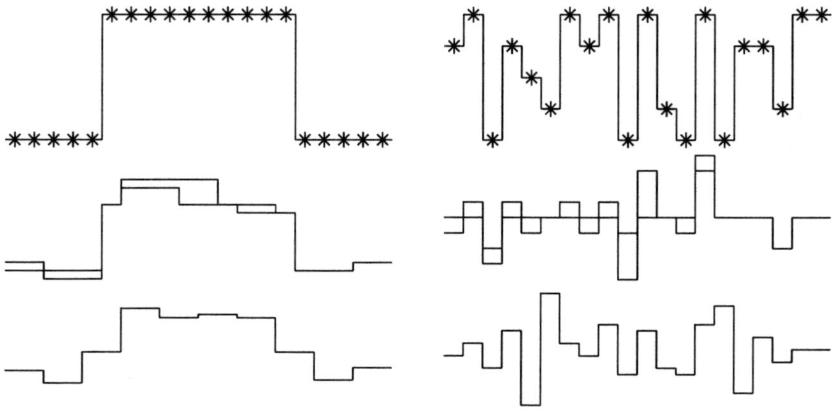

Figure 7.9. One stage of decomposition by L_1U_1-, U_1L_1- and Haar-decomposition.

Having suggested and heuristically argued many "advantages" the $LULU$-decompositions have, it would be appropriate to derive some theorems that can support some of these known, observed or believed properties of the $LULU$-decomposition.

In linear orthogonal decompositions the squares of 2-norm are preserved. A pure n-pulse has the same "energy" as n 1-pulses of the same amplitude, but, depending upon how these 1-pulses are distributed, can have a total variation between 2α and $2n\alpha$. The total variation is therefore a measure of the "resolution" of features. Pursuing this idea leads to the following substantial results.

The operators involved were originally intended for nonlinear smoothing, or pre-smoothing for the removal of impulsive noise. On attempting to clarify and quantify some observed experimental observations and case studies, it is natural to choose total variation of a sequence as a measure of smoothness.

A wavelet decomposition of a sequence $x \in \ell_p$ with order 1 or 2 spline wavelets will eventually lead to a constant sequence, which has to be the null sequence. Thus all the "energy" has been peeled off into the (wavelet-) frequency layers, since the squares of the 2-norms of Px and $x - Px$ add up to the square of the 2-norm of x, so that

$$||x - P_1x||_2^2 + ||P_1x - P_2P_1x||_2^2 + \cdots = ||x||_2^2.$$

The behavior of the total variation of x and all the components of the decomposition at the intermediate stages is of interest, if smoothing is desired for the purpose of exposing significant features in the sequence x without too much damage. In a wavelet-decomposition, as in a Fourier decomposition, it is natural to stop the decomposition process when the frequency layers do not contain significant energy any more. Similarly it would be convenient if the reduction in variation has some natural measure giving precise indication of what fraction of the total variation has been removed. We assume that the features to be exposed by smoothing are given by a sequence e to which higher resolution noise, given by a sequence r, has been added to produce a significant increase in variation. Since $T(x) = T(e+r) \leq T(e) + T(r)$ in general we are therefore assuming that $T(x) \approx T(e)+T(r)$. We should like to stop smoothing when the total variation goes significantly below that of $T(e)$, which is not known but perhaps estimated. If the noise r is of much higher "resolution" than the desired features of the sequence, the successive peeling off of resolution layers could be expected to decrease the variation steadily until most of the noisy features are removed. Reaching the expected resolution level of e should result in a further strong reduction, perhaps signaling that the unknown resolution level of e has been reached, and further "smoothing" would partially erase these. This Heuristic motivation could be experimentally supported, but clearly the underlying assumption of a "proportional" allocation of the total variation to each resolution layer, is crucial in the appropriateness.

The $LULU$-decompositions have such a remarkable property. It can be developed in the following way.

Lemma. Let x be $(n-1)$-monotone. If j is a point where $U_n x_j \neq x_j$ and $x_{j-1} \neq x_j$, then $U_n x_j = \min\{x_{j-1}, x_{j+n}\}$ and $U_n x_{j-1} = x_{j-1}$.

Proof. If $U_n x_j$ differs from x_j it must be larger, since $U_n x \geq x$.

$$U_n x_j = \min\{\max\{x_{j-n}, \ldots, x_j\}, \ldots, \max\{x_j, \ldots, x_{j+n}\}\} > x_j,$$

implies that each of the maxima is larger that x_j, so that there are at least two values $x_\ell \in \{x_{j-n}, \ldots, x_j\}$ and $x_r \in \{x_j, \ldots, x_{j+n}\}$, such that $x_\ell, x_r > x_j$. But x is $(n-1)$-monotone, so that each set of n successive elements of x are monotone. Then

$$x_{j-n} \geq \cdots \geq x_\ell \geq \cdots x_{j-1} \geq x_j \leq x_{j+1} \leq \cdots \leq x_r \leq \cdots \leq x_{j+n}.$$

Noting $x_{j-1} \neq x_j$, from the assumption of the lemma, it must follow that $x_{j-1} > x_j$. From this it follows that there must be a constant section of equal values, since $x_{j-1} > x_j \geq \cdots \geq x_{j+n-1}$ and $x_j \leq \cdots \leq x_{j+n-1} \leq x_{j+n}$, imply that $x_j = x_{j+1} = \cdots = x_{j+n-1}$.

At least one of the points $x_{j+1}, \ldots, x_{j+n}$ must be strictly larger than x_j, so that this value must be x_{j+n}. Therefore also $U_n x_{j+n} = x_{j+n} = x_r$ and $r = j+n$. Thus $U_n x_j = \min\{\max\{x_{j-1}, \ldots, x_{j+n-1}\}, x_{j+n}\} = \min\{x_{j-1}, x_{j+n}\}$. Furthermore, clearly $U_n x_{j-1} = x_{j-1}$, since $\max\{x_{j-1}, x_j, \ldots, x_{j-1}\} = x_{j-1}$. □

Although a more general theorem was previously derived, it is instructive to prove the following theorem again, so that the features of the sequences that are separated become clearer in the case of successive separation into resolution levels.

Theorem 7.1. For $x \in \mathcal{M}_{n-1}$, $n \geq 1$,

$$T(x) = T(\bigvee^{n} x) + T(U_n x - x) = T(\bigwedge^{n} x) + T(L_n x - x).$$

Proof. Consider the sequence $t = \{t_j\}$ of integers where $U_n x_{t_j} > x_{t_j}$ and $x_{t_j+1} \neq x_{t_j}$. (Recall the previous lemma, and note that $\{x_{t_j-n}, \ldots, x_{t_j}\}$ are $(n-1)$-monotone, $x_{t_j-1} > x_{t_j}$.) The sequence t cannot contain two consecutive integers, so that

$$T(x) = \sum_{j=-\infty}^{\infty} \tau_j, \quad \text{with} \quad \tau_j = \sum_{i=t_j-n}^{t_{j+1}-n-1} |x_{i+1} - x_i|.$$

Consider a specific j, with $k = t_j$ and $m = t_{j+1}$, and for notational convenience, let $c_{j+1} = w$ be the first index after k such that $x_w > x_{w+1}$. Clearly w exists in $[k+1, m-1]$. Since x is $(n-1)$-monotone, and therefore

$$x_{k-1} > x_k \geq x_{k+1} \geq \cdots \geq x_{k+n-1} \quad \text{and} \quad x_k \leq x_{k+1} \leq \cdots \leq x_{k+n},$$

it follows that

$$x_k = x_{k+1} = \cdots = x_{k+n-1}.$$

A similar argument yields $x_{w-n} = \cdots = x_{w-1} = x_w$.

By the previous lemma $Ux_k = \min\{x_{k-1}, x_{k+n}\}$ implies that x_{k-1}, x_{k+n} and w must be larger than x_k. Since t_{j+1} is the first integer after t_j where the sequence changes from a monotone decreasing to monotone increasing set of values, $\{x_{k+n}, \ldots, x_{m-n-1}\}$ has the following structure; $\{x_{k-n}, \ldots, x_{k-1}\}$ is a monotone decreasing section, $x_{k-1} > x_k = x_{k+1} = \cdots = x_{k+n-1}$, $\{x_k, \ldots, x_w\}$ is monotone increasing and $x_w > x_{w+1} \geq \cdots \geq x_m$. Let:

$$\begin{aligned}
\mu_j &= \sum_{i=k-n}^{m-n-1} |\bigvee^{n} x_{i+1} - \bigvee^{n} x_i| \\
&= \sum_{i=k-n}^{k-2} |\bigvee^{n} x_{i+1} - \bigvee^{n} x_i| + \sum_{i=k-1}^{k-1} |\bigvee^{n} x_{i+1} - \bigvee^{n} x_i| + \sum_{i=k}^{m-n-1} |\bigvee^{n} x_{i+1} - \bigvee^{n} x_i|.
\end{aligned}$$

The first of these three sums is equal to

$$\sum_{i=k-n}^{k-2} |x_{i+1} - x_i|,$$

since the set $\{x_{k-n}, \ldots, x_{k+n-1}\}$ is monotone decreasing.

The next sum is $|\bigvee^n x_k - \bigvee^n x_{k-1}| = |x_{k+n} - x_{k-1}|$, and the last sum has two cases:

(i) Suppose $w > m-n-1$. Then

$$\sum_{i=k}^{m-n-1} |\bigvee^{n} x_{i+1} - \bigvee^{n} x_i| = |x_{k+n+1} - x_{k+n}| + \cdots |x_{w-n} + x_{w-n-1}|,$$

since $m-n-1 > w-n-1$ if $w < m$ and $x_{w-n} = x_{w-n+1} = \cdots = x_{w-1} = x_w$. This is the variation on a monotone increasing section and yields $|x_w - x_{k+n}|$. Thus

$$\begin{aligned}\mu_j &= \sum_{i=k-n}^{k-2} |x_{i+1} - x_i| + |x_{k+n} - x_{k-1}| + x_w - x_{k+n} \\ &= \sum_{i=k-n}^{k-1} |x_{i+1} - x_i| - |x_k - x_{k-1}| + |x_{k+n} - x_{k-1}|.\end{aligned}$$

But

$$\begin{aligned}x_w - x_{k+n} = x_w - x_{k+n-1} + x_{k+n-1} - x_k &= \sum_{i=k}^{m-n-1} |x_{i+1} - x_i| + x_{k+n-1} - x_k \\ = \tau_j + (x_k - x_{k-1}) + |x_{k+n} - x_{k-1}| + x_k - x_{k+n}.\end{aligned}$$

Therefore

$$\begin{aligned}\mu_j &= \tau_j + 2x_k - 2\min\{x_{k+n}, x_{k-1}\} \\ &= \tau_j - 2(U_n x_k - x_k), \quad \text{from the previous lemma} \\ &= \tau_j - \frac{2}{m} \sum_{i=k-n}^{m-n-1} |U_n x_i - x_i| = \tau_j + \sum_{i=k-n}^{m-n-1} |U_n x_{i+1} - x_{i+1} - U_n x_i + x_i|.\end{aligned}$$

The last equality comes from the fact that $Ux_i - x_i = Ux_k - x_k$ for $i = k, \ldots, k+n-1$ and zero elsewhere in the interval $[k-n, \ldots, m-n-1]$, and then the variation of this block pulse is simply twice the height.

(ii) Suppose $m-n-1 \geq w$. Then

$$\sum_{i=k}^{m-n-1} |\bigvee^{n} x_{i+1} - \bigvee^{n} x_i| = x_w - x_{k+n} + x_w - x_{m-n} \quad \text{or}$$

$$\sum_{i=k}^{m-n-1} |\bigvee^{n} x_{i+1} - \bigvee^{n} x_i| = x_w - x_k + x_w - x_{m-n} + x_k - x_{k+n}.$$

This again yields

$$\mu_j = \tau_j - \frac{2}{n} \sum_{i=k-n}^{m-n-1} |U_n x_i - x_i| = \tau_j - 2(U_n x_k - x_k).$$

In both cases therefore

$$
\begin{aligned}
T(\bigvee^n x) &= \sum_{j=-\infty}^{\infty} \mu_j = \sum_{j=-\infty}^{\infty} \tau_j - \sum_{j=-\infty}^{\infty} \frac{2}{n} \sum_{i=t_j-n}^{t_{j+1}-n-1} |U_n x_i - x_i| \\
&= T(x) - T(U_n x - x) \\
\text{or} \quad T(x) &= T(\bigvee^n x) + T(x - U_n x).
\end{aligned}
$$

By a similar argument, or by the usual duality argument, $T(x) = T(\bigwedge^n x) + T(x - L_n x)$. This proves the theorem. As was proved previously, a corollary follows easily: □

Corollary: For $x \in \mathcal{M}_{n-1}, n \geq 1$:

$$
\begin{aligned}
T(x) &= T(U_n x) + T(x - U_n x) \quad \text{and} \\
T(x) &= T(L_n x) + T(x - L_n x), \\
T(x) &= T(L_n U_n x) + T(x - L_n U_n x) \quad \text{and} \\
T(x) &= T(U_n L_n x) + T(x - L_n U_n x).
\end{aligned}
$$

Theorem 7.2. Then $||U_n x - L_n x||_1 = ||U_n x - x||_1 + ||x - L_n x||_1$ for $n \geq 0$.

Proof. Since $U_n x_i \geq x_i \geq L_n x_i$ it follows easily that

$$
\begin{aligned}
\sum_{i=-\infty}^{\infty} |U_n x_i - L_n x_i| &= \sum_{i=-\infty}^{\infty} U_n x_i - L_n x_i \\
&= \sum_{i=-\infty}^{\infty} (U_n x_i - x_i + x_i - L_n x_i) \\
&= \sum_{i=-\infty}^{\infty} U_n x_i - x_i + \sum_{i=-\infty}^{\infty} x_i - L_n x_i \\
&= ||U_n x - x||_1 + ||x - L_n x||_1.
\end{aligned}
$$
□

In the above comparisons the complications at endpoints of a finite sequence are avoided for the sake of simplicity. These problems can usually be treated satisfactorily in a variety of different ways by letting both beginning and end of a finite sequence move to zero in a variety of different reasonable ways. Furthermore, there is an arguably unfair advantage that is easily overlooked if only the first decompositions are considered.

If the sequences are finite, then a Haar-decomposition projects the sequence x onto a nested set of subspaces, each being of half the dimension of the previous until a constant is reached. The $LULU$-operators map, in a way that is projection-like, (roughly speaking, as near to a projection as can be expected with a nonlinear operator) onto nested subsets $\mathcal{M}_1 \supset \mathcal{M}_2 \supset \mathcal{M}_3 \supset \cdots$ etc. Since these are not subspaces, there is no question of dimension, but to be fair it should be realized that there is not a decomposition into octaves, but rather a linear decrease in "flexibility" (for lack of a better word). A fairer comparison would be if the decomposition were from $\mathcal{M}_0$ to $\mathcal{M}_1$ to $\mathcal{M}_3$ to $\mathcal{M}_5$ etc., since, if the range of the projections P_j of the Haar-decomposition is R_j we have that $R_j \subset \mathcal{M}_{2^j-1}$. The proof of this is obvious if we observe that a sequence in R_j is a sampled B-spline of order 1, which is $(2^j - 1)$-monotone, since it is a sampling of a piecewise constant function. The sequence is therefore made up of successive sections of 2^j equal values, and clearly every $2^j + 1$ successive values are monotone.

At this stage there has been no attempt to suggest comparison in a significant advantage of the wavelet-decomposition, namely the economizing in representation and computation by the fact that the smoother component and the wavelet component can be economically stored by the respective spline and wavelet basis. This yields an effective representation in no more numbers than the original number of elements of the sequence. These coefficients can furthermore be quantized without major distortion in the reconstructed sequence.

In the $LULU$-decomposition there are no bases that are generally useful for economization. There are several possibilities for savings by coding the noise components, which can be progressively more sparse. Similarly the smoother part could have progressively larger constant sections permitting some economizing. This whole issue is not addressed here, and it is generally complicated, as in the case of two-dimensional image processing, with both wavelet decompositions and Median decompositions. The primary comparisons to be made here are more fundamental in nature. The $LULU$-decompositions are an alternative to the prevalent median transform of the same type. These seem generally to be good in the two-dimensional case of image processing (3). A disadvantage listed is the computational complexity. A more serious disadvantage seems to be the lack of theory. Wavelet theory has a comprehensive and beautiful theory for analysis. For linear operators, Fourier Transforms and their inverses provide a framework for analysis. For a large class of operators, Mathematical Morphology provides a framework of analysis, and is well developed, especially in the two-dimensional case. Starck, Murtagh and Bijaoni state that median transforms are considered better than Morphological Filters. This may be when considering only the simpler compositions like U_n and L_n, in the one-dimensional case, when $L_n \leq M_n \leq U_n$ is true, but the interval $[L_n, U_n]$ is too large and L_n and U_n are only comparable to M_n in approximating properties when the significant features in the noise are one-sided. The inequality $U_nL_n \leq M_n \leq L_nU_n$ is much sharper and the operators U_nL_n and L_nU_n are good approximations to the Median. This is visible in the foregoing examples where the first decompositions with L_nU_n and U_nL_n are compared. Since this difference can be argued to be within the fundamental uncertainty interval of the concept "impulse" and an associated concept of "resolution", they must both be expected to be equivalent to the Median for the purpose at hand.

The advantage argued here is in the application to analysis of measurements. Significant features of measurements that are fundamental are "edges", "local trend", and "pulses". The words "edges", "trend", "impulse", "pulse" and "resolution" are all widely used in science and technology and are very often not clearly defined in the context. For a multiresolution decomposition to be called such, it may be prudent to become somewhat more precise.

In chemistry, for example, a spectroscopy apparatus can be said to have a resolution sufficient to distinguish between spectral lines of two chemical compounds. These may be of Gaussian-shape and the instrument must be able to separate two definite maxima, even in the presence of some reasonable expected noise. If the instrument has a higher resolution it can separate a spectrum of each of these compounds into several Gaussian pulses from the presence of a few ionisation species in the compounds. When the instrument is even better it could even break up these pulses into pulses of different isotopes in the species. It is instructive to consider again an ambiguity intrinsic in the concept of a pulse with a simple example.

In Figure 7.10, a pair of level 1 pulses is seen to be ambiguously interpretable as a pulse at level three with two level 1 pulses superimposed, or as two level 1 impulses. Clearly alternate interpretations are also possible, but the ambiguity is at least apparent from the different sequences $x - L_1U_1x$ and $x - U_1L_1x$.

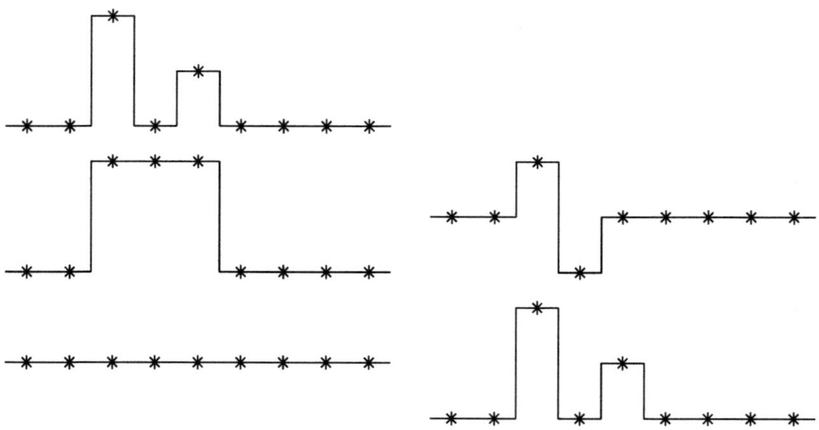

Figure 7.10. The first decompositions of a two pulse signal by L_1U_1 and U_1L_1.

After the next two decompositions no pulse of lower resolution will be present. The only activity is in the first three layers indicating the fundamental resolution interval $[1, 3]$. The Haar decomposition of the same pulse in Figure 7.11 smears activity into all levels, making a stricter use of the term "resolution" difficult.

It is not the purpose here to expand on all the various advantages over wavelet transforms, but to indicate the type of use that has been, and can be made in the detection of detail at various resolution levels. The method of analysis is

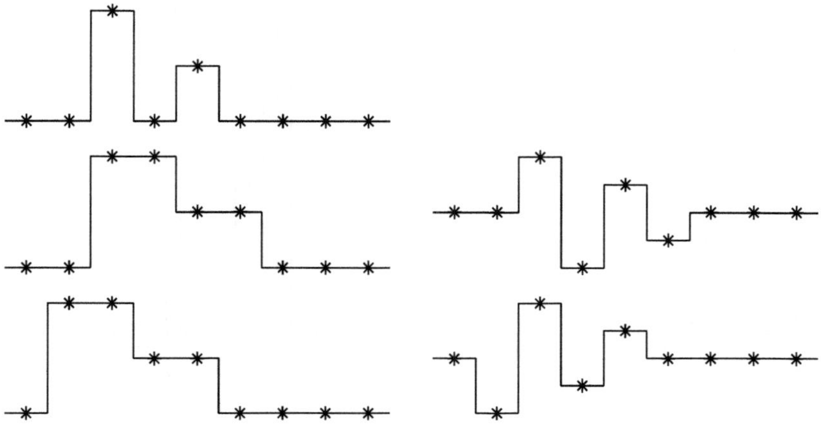

Figure 7.11

relatively unknown, and based on the strong mathematical structure on the basic operators. Having defined a pure n-impulse p_n it may be useful to define a pulse p of resolution between m and n if $p_m \leq p \leq p_n$. If superimposed noise is comparatively small, it is clear that, by the syntoneness of the operators involved in the $LULU$-decomposition, the pulse will appear in the mth level and disappear in the nth level. Thus it may be a more precise way of defining what is meant by a pulse at a resolution level (or interval). If two such pulses are separated sufficiently they can be resolved. When they are superimposed, or close enough, the fundamental uncertainty at that resolution level can be recognized by the difference between the decomposition with L_nU_n and U_nL_n.

At each level n the operators L_n, U_n, U_nL_n and L_nU_n are all idempotent and co-idempotent and all compositions are one of these four. This near-ring of operators has a complete order given by $L_n \leq U_nL_n \leq L_nU_n \leq U_n$. In the decomposition procedure several layers of such operators are used, and since they are all syntone operators, relative orders are inherited and remain useful for analysis. The well-established Median operators, popularly used for smoothing out impulsive noise and accepted to be very good for preservation of edges and other significant features, are contained between such $LULU$-outputs. When decompositions with L_nU_n and U_nL_n are compared, the difference indicates an "amount of ambiguity, which turns out to be very useful. Using the "commutators" like $L_nU_n - U_nL_n$ resulted in the design of recording instruments for times of arrival of shock pulses for location purposes. In the first application attempted, accuracy of such time of arrival estimates have exceeded the best previous designs of an international company by a large margin. Figure 7.12 below shows the histogram of arrival errors of the $LULU$-based trigger compared to the triggers then in use. The advantage seems due to the strong edge preservation (shape preservation) in the presence of large quantities of noise.

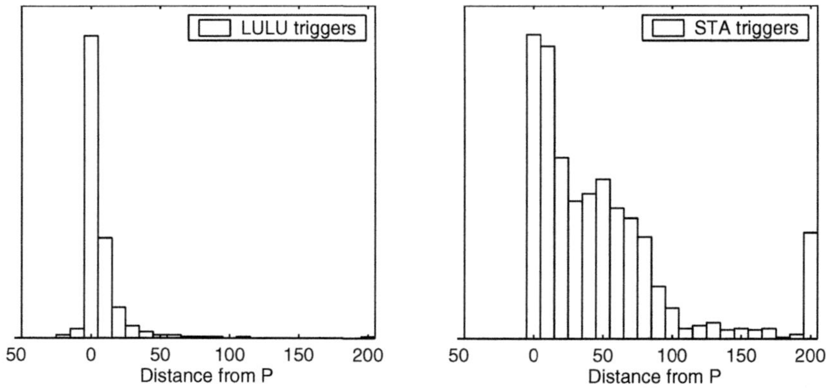

Figure 7.12. Comparison of time-of-arrival errors between $LULU$-estimates and the current estimates based on ratios of linear smoothers, in histogram form.

At this stage the merit of an alternative Multiresolution Analysis is sufficiently apparent, but the fundamental problem that looms over the whole idea is the nonlinearity of the decomposition operators (separators). If we have a Wavelet (or Fourier-) decomposition we have a simple change of basis and any linear combination of the basis vectors are guaranteed to be decomposed into precisely that set of sequences again.

Suppose however that a sequence $x \in M_0$ has been decomposed into a sequence of resolution sequences $r^{(n)}$; $n = 1, 2, \ldots$ by a Multiresolution Decomposition with the operators L_nU_n, for $n = 1, 2, \ldots$. Several questions arise immediately:

1. Will a basic modification of any resolution level, such as multiplication by a non-negative factor, have a decomposition that will have simply that resolution level similarly modified, but otherwise identical?
2. What is to be expected when several high, or low or arbitrary, resolution sequences undergo basic modification and are added together and decomposed?

Such questions need to be resolved before any degree of confidence can be attached to the value of the whole decomposition procedure.

8. The Discrete Pulse Transform

Bleib nicht auf ebnem Feld!
Steig nicht zu hoch hinaus!
Am schönsten sieht die Welt
von halber Höhe aus.

Nietzsche

To address the question of consistency of the DPT, it is convenient to introduce a notation and compare with the Direct Fourier Transform (DFT or FFT). As in the case of the Fourier Transform, we often seek a "band pass filter". The information required needs to have the high resolution detail and the low resolution drift removed. What is the distortion that results? In the DFT the removal of high frequencies results in smearing of edges (detail), and if the trend (drift) is not removed there will be unwanted distortion at the edges. Consistency of the DPT seems to be unobtainable, since we have compositions of essentially nonlinear operators at work. The remarkable shape preservation properties of the basic operators do however keep sufficient hope alive to try the seemingly impossible.

Conceptually a sequence $x = \langle x_i; i = 0, 1, 2, \ldots, N = 2^m$, with $x_N = x_0\rangle$ is decomposed into $\frac{N}{2} + 1$ sequences $s^{0)}, s^{(1)}, \ldots, s^{(\frac{N}{2})}$, where $s^{(0)}$ is a constant sequence and $s^{(n)}$ is a periodic sequence with frequency n times the fundamental frequency. As is well known, the mappings D_n that map x onto $s^{(n)}$ are projections onto a one-dimensional space when $n = 0$ and a two-dimensional space otherwise. Each component sequence can thus be associated with two numbers, either coefficients with respect to a chosen basis or an amplitude and phase pair. The notation $D(x) = DFT(x) = [D_0(x), D_1(x), \ldots, D_M(x)]$, with $M = 2^{m-1}$, is convenient, with $s^{(i)} = D_i(x)$. Since the process is essentially a basis transformation it is obvious that the DFT is component consistent in the following sense;

$$\text{If} \quad z = \sum_{i=0}^{M} \alpha_i D_i(x), \quad \text{then} \quad D_n(z) = \alpha_n D_n(x), \quad \text{for each} \quad n = 0, 1, \ldots, M.$$

Furthermore, the orthonormality yields the "Energy" preservation law;

$$||x||_2^2 = \sum_{i=0}^{M} ||D_i(x)||_2^2,$$

so that it is possible to allocate a percentage of the energy in x to each frequency level.

The proposed DPT (Discrete Pulse Transform), can then conveniently be viewed as the mapping of a sequence $x = \langle x_i; i = 0, 1, \ldots, N = 2^m, \quad \text{with} \quad x_N = x_0 \rangle$ onto a vector of sequences $r^{(n)} = D_n(x)$, at different "resolution levels", such that

$$D(x) = DPT(x) = [D_1(x), D_2(x), \ldots, D_N(x), D_0(x)], \quad \text{with} \quad D_0(x) = C_N x.$$

There are two equivalent natural primary choices for such decomposition procedures, based on the smoothers L_nU_n, or U_nL_n. Considering the first choice, the decomposition proceeds recursively. The sequence x is separated by the separator L_1U_1 into a "smoother" sequence L_1U_1x and the highest resolution sequence $(I - L_1U_1)x = D_1(x) = r^{(1)}$. The smoother part L_1U_1x is then separated by L_2U_2 to yield the second "resolution component" $r^{(2)} = D_2(x) = (I - L_2U_2)L_1U_1x$ and the smoother part $L_2U_2L_1U_1x$.

This is continued until after the Nth separation only a constant sequence $D_0(x)$ remains. If the restriction $x_N = x_0$ on x is omitted, the "lowest resolution" component $D_0(x)$ will be monotone, but not necessarily constant. We shall however demand $x_N = x_0$, as this is also done with the DFT where "trend" is first removed. Clearly the properties of the DPT will result from the properties of the smoothers L_nU_n. The DPT is not a basis transformation like the DFT, it is not even linear in any of its separators. It is therefore not even clear that, if we take an individual resolution component $r^{(n)}$ of x, the decomposition of the sequence $r^{(n)}$ will yield $r^{(n)}$ at level n, or that the other resolution levels would have zero sequences. More generally it can be stated (any example is a proof) that the decomposition of a sum is not the sum of the decompositions nor is $DPT(\alpha x) = \alpha DPT(x)$, for all α.

Furthermore the LU decomposition does not even confirm that of UL, in the sense that if the nth resolution component $r^{(n)}$ of x, as decomposed by the operators LU, is decomposed by the UL decomposition, then in general $r^{(n)}$ will not be the nth resolution component of x. The consistency of each DPT will therefore require careful analysis. As it turns out a remarkable consistency exists, which needs to be proved carefully.

The consistency of the individual separators L_nU_n (or U_nL_n) is the key to initial consistency analysis. It is worth reviewing the argument here. If an operator S is to separate a sequence into a "signal" Sx and (additive-) "noise" $(I - S)x$ then this should ideally be consistent. A projection is an example of a consistent "separator" in that having obtained Sx and $(I - S)x$ from a given sequence, separation of Sx would yield only $S(Sx) = Sx$ and $(I - S)Sx = 0$, which signifies no noise. Similarly if $(I - S)x$ is separated $(I - S)((I - S)x) = (I - S)x$, as a projection is linear and $S(I-S)x = 0$, which signifies pure "noise" and no "signal".

If an operator is not linear, then idempotence results in consistent signal extraction as then $S(Sx) = Sx$, but $S(I-S)x$ is not necessarily the zero sequence. For consistent separation both idempotence and co-idempotence have to be required.

Such operators are separators, and were shown to have only two eigenvalues 1 and 0, corresponding to the "signals" and "noise" as eigensequences respectively and $\mathcal{M}_n$ is exactly the set of eigensequences with respect to 1 of L_nU_n and U_nL_n, but they do not form subspaces, since the sum of two signals need not be a signal. (As was shown, the operators $I - L_nU_n$ and $I - U_nL_n$ do not share their eigensequences w.r.t. the eigenvalue 0. L_nU_n and U_nL_n do not always agree on what is "noise", although they agree on what is "signal", since $L_nU_n(U_nL_n) = U_nL_n$ and $U_nL_n(L_nU_n) = L_nU_n$.).

If the DPT with L_nU_n or the DPT with U_nL_n are considered they can be viewed as separating a given sequence x into $N+1$ component sequences at different resolution levels. Primary consistency demands that if any resolution level is taken through the separation process, it should yield zero sequences everywhere else and the full sequence at the appropriate resolution level. The following theorem guarantees just that.

Theorem 8.1. Let x be a given sequence in $\mathcal{M}_0$ and

$$DPT(x) = [D_1x, D_2x, \ldots, D_Nx, D_0x].$$

Then $D_j(D_ix) = \delta_{ij}D_ix$, where δ_{ij} is the Kronecker delta.

Proof. Consider the DPT with L_nU_n as separators. Firstly $D_1x = (I - L_1U_1)x$ and

$$\begin{aligned}(I - L_1U_1)D_1x &= (I - L_1U_1)((I - L_1U_1)x) = (I - L_1U_1)^2x \\ &= (I - L_1U_1)x,\end{aligned}$$

by the co-idempotency of L_1U_1. Thus $D_1(D_1x) = D_n(x)$, and since $L_1U_1(I - L_1U_1) = 0$ the zero sequence is passed to the next separators and yields the zero sequence at each resolution level.

For $n > 1$ the sequence $D_nx = (I - L_nU_n)C_{n-1}x$, which is in M_{n-1}, passes unchanged through the smoothers L_jU_j and clearly yields zero sequences $r^{(j)} = 0$ for $j \leq n-1$. At the separator L_nU_n the output 0 is passed to the next separator yielding subsequent zeroes, since

$$(I - L_nU_n)D_nx = (I - L_nU_n)(I - L_nU_n)C_{n-1}x = (I - L_nU_n)C_{n-1}x = D_nx.$$

Thus D_nx appears in full at resolution level n, and $D_j(D_nx) = \delta_{jn}D_nx$.

A similar proof holds for the DPT with U_nL_n. □

Thus individual resolution components of a sequence are consistently decomposed and, as a next stage, it is reasonable to investigate the consistency as a "low-pass" smoother. A sequence z is formed by omitting the first (highest-) resolution components of a sequence x. The following theorem demonstrates such consistent decomposition of z.

Theorem 8.2. Let $x \in$ in $\mathcal{M}_0$ that is decomposed by the DPT of Theorem 8.1, and

$$z = x - \sum_{i=1}^{n} r^{(i)}, \quad \text{with} \quad r(i) = D_ix. \quad \text{Then} \quad D_m(z) = \begin{cases} 0 & \text{for } m < n, \\ r^{(m)} & \text{for } m \geq n \end{cases}.$$

Proof. By induction on n,

$$z = x - \sum_{i=1}^{n} r^{(i)} = C_n x.$$

Since $C_n x \in M_n$ the smoothers $L_m U_m$ leave $C_n x$ unchanged while $m < n$. Thus all the high resolution components of z are zero until smoother $L_{n+1}U_{n+1}$ is reached, which separates $z = C_n x$ into $C_{n+1}x$ and the resolution component $D_{n+1}z = (I - L_{n+1}U_{n+1})C_n x = D_{n+1}x$. The rest of the decomposition clearly follows as for x, yielding the same values for the low resolution components.

A similar proof holds for the DPT with $U_n L_n$. □

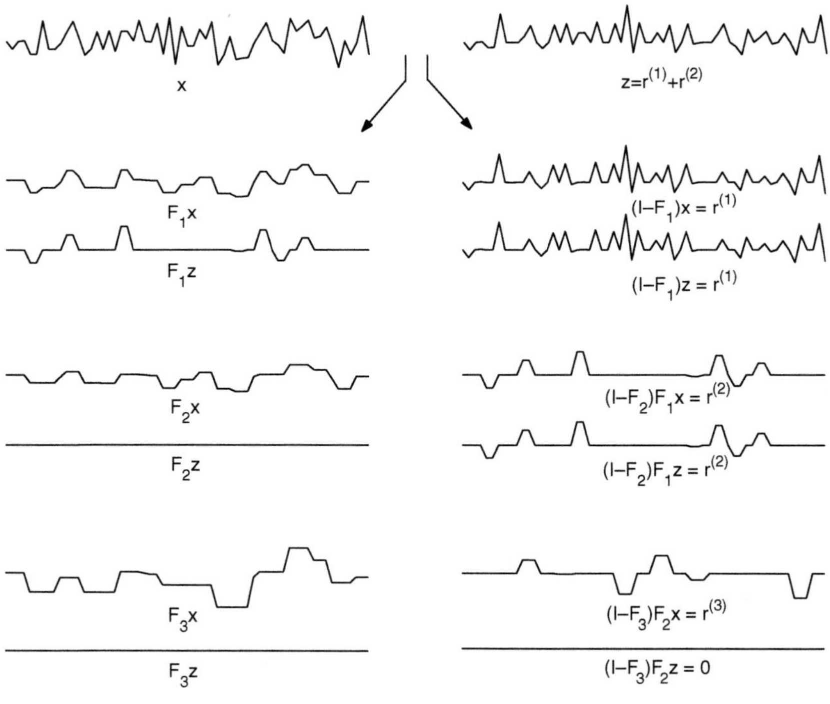

Figure 8.1. Theorem 8.2 verified by a random sequence.

A further level of consistency would be to omit the low resolution components , as would be the case in a "high-pass" smoother. Here however the result requires some additional theory. Suppose, for instance that x has only two nonzero resolution levels $r^{(1)} = D_1 x$ and $r^{(2)} = D_2 x$. By the previous theorems, omitting any of the resolution levels to form a sequence z, the decomposition would be consistent with that of x. But if there are three nonzero resolution levels, there are two partial sums that are problematic.

Assume $z = r^{(1)} + r^{(2)} = x - r^{(3)}$. $D_1 z = (I - L_1U_1)(I - L_1U_1)x + (I - L_2U_2)(C_1 x)$, and since neither L_1U_1 nor $I - L_1U$ are linear, it is not clear that $D_1 z = D_1 x = r^{(1)}$. The result seems to hinge on the fact that L_1U_1 acts linearly on the particular sum $r^{(1)} + r^{(2)} = x - C_2 x = (I - C_2)x$. Omitting the resolution component $r^{(2)}$ yields similar problems in the proof of the consistent decomposition of the partial sum. Since experimentation suggests consistency of this type, there must exist some theory not yet explored. The consistency seems too good to be true. We start with the "band pass" case.

Theorem 8.3. Let $z = r^{(m)} + r^{(m+1)} + \cdots + r^{(n)}$. Then z is decomposed consistently.

Proof. For $j < m$, and noting that $C_{m-1}(x)$ is $m-1$ monotone and $z = (I - C_n)C_{m-1}(x)$,

$$\begin{aligned} C_j z &= C_j(I - C_n)C_{m-1}(x) = (C_j - C_n)C_{m-1}(x), \quad \text{by Theorem 6.18.} \\ &= (I - C_n)C_j C_{m-1}(x) = (I - C_n)C_{m-1}(x) = z. \end{aligned}$$

The sequence z is therefore passed unaltered through the first $m-1$ separators and means that $D_j(z) = 0$, for $j < m$.

For $m \leq j \leq n$; $\quad C_j(z) = (I - C_n)C_j C_{m-1}x = (I - C_n)C_j(x)$ and

$$\begin{aligned} D_j(z) &= (C_{j-1} - C_j)z = (I - C_n)C_{j-1}x - (I - C_n)C_j x \\ &= C_{j-1}x - C_j x - C_n C_{j-1}x + C_n C_j x = C_{n-j}x - C_j x = D_j(x). \end{aligned}$$

Finally $C_n(z) = C_n(I - C_n)C_{m-1}x = 0$, by the co-idempotence of C_n, or $C_n(I - C_n) = 0$. Thus the components of z come out fully consistently. □

Clearly the DPT is consistent as a "high pass" smoother. (The case where $m = 1$) and as a "band pass" smoother. (General $m < n$). A further result, demonstrating further linear behavior under specific conditions is required to prove additional consistency. This result comes from Theorem 6.19 by the following remarkably general theorem.

Theorem 8.4. Let $x \in \mathcal{M}_0$ and $DPT(x) = [D_1x, D_2x, \ldots, D_Nx, D_0x]$, with $D_0 x = C_N x$. If $\alpha_i \geq 0$, then $z = \sum_{i=1}^{n} \alpha_i D_i x$ is decomposed consistently, for $n \leq N$.

Proof. Let $z = \sum_{i=1}^{n} \alpha_i D_i x$. Noting that $D_i = (1 - C_i)C_{i-1}$ are all ftp we see that

$$\sum_{i=m}^{n} \alpha_i D_i x = \left(\sum_{i=m}^{n} \alpha_i (I - C_i) C_{i-1} \right) x = A_m C_{m-1} x \in \mathcal{M}_{m-1},$$

since $C_{i-1} = C_{i-1}C_m$ for $i \leq m$, where

$$A_m = \left(\sum_{i=m}^{n} \alpha_i (I - C_i) C_{i-1} \right)$$

is ntp, since it is a convex combination of ftp operators.

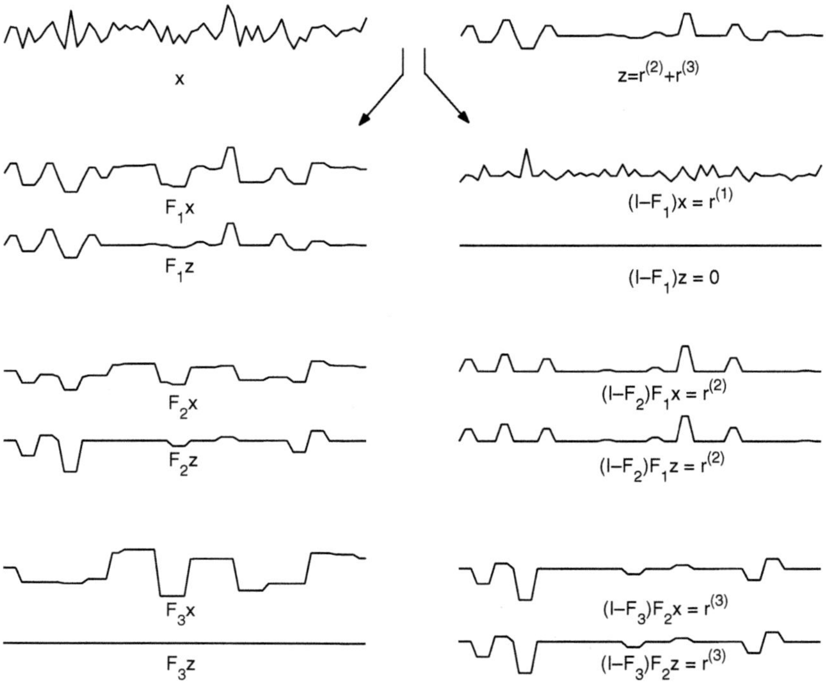

Figure 8.2. Theorem 8.3 verified by a random sequence.

Assume that

$$C_{j-1}z = \left(\sum_{i=j}^{n} \alpha_i D_i\right) C_{j-1}x \quad \text{for} \quad j < n.$$

(Clearly it is true for $j = 1$).

$$\begin{aligned}
C_j z &= L_j U_j(\alpha_j D_j + \left(\sum_{i=j+1}^{n} \alpha_i D_i\right) C_{j-1}x \\
&= L_j(\alpha_j U_j D_j + A_{j+1}C_j)C_{j-1}x, \quad \text{by Theorem 6.19} \\
&= L_j(\alpha_j U_j(I - L_j U_j) + A_{j+1}L_j U_j)C_{j-1}x \\
&= L_j(\alpha_j(I - L_j)U_j + A_{j+1}L_j U_j)C_{j-1}x, \quad \text{by Theorem 6.15} \\
&= L_j(\alpha_j(I - L_j) + A_{j+1}L_j)U_j C_{j-1}x \\
&= (\alpha_i L_j(I - L_j) + A_{j+1}L_j)U_j C_{j-1}x \\
&= 0 + A_{j+1}C_j x.
\end{aligned}$$

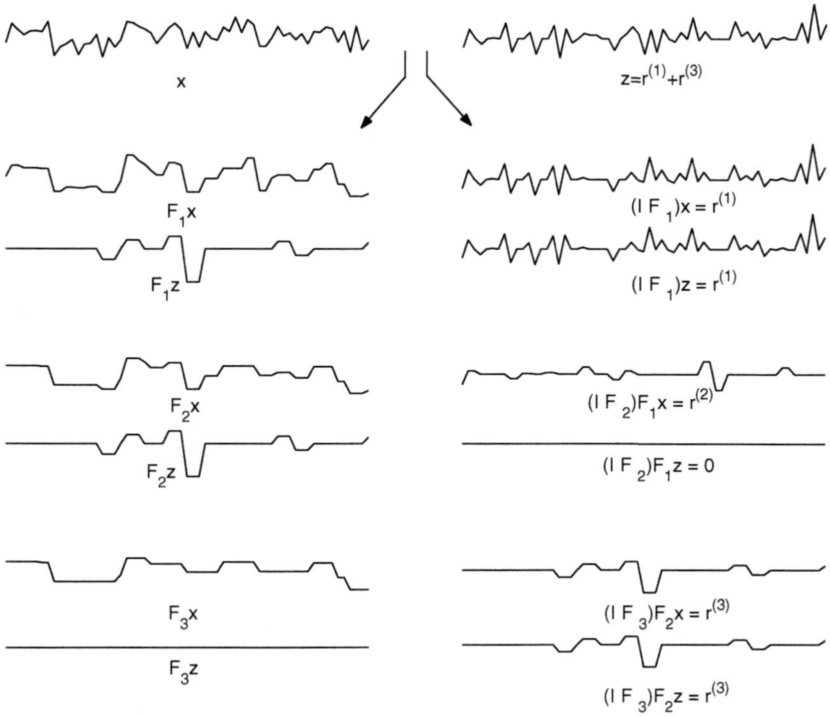

Figure 8.3. Omitting more general selected resolution levels.

Thus

$$D_j = (C_{j-1} - C_j)z = \alpha_j D_j x \quad \text{and} \quad C_j z = \left(\sum_{i=j+1}^{n} \alpha_i D_i \right) C_j x.$$

A standard induction argument completes the proof of the theorem. □

Corollary. Any sequence $\sum_{i=0}^{N} \alpha_i D_i x$, with $\alpha_i \geq 0$ is decomposed consistently.

The basic alternative DPT using the operators $U_n L_n$ has a similar consistency and this can be proved in the same way as above. What is important to note is that they generally give different resolution components. When the sequence is strongly correlated, the two decompositions yield essentially similar results, but the addition of additive random noise yields some separation, proportional to the amplitude of this noise. What is important is that the two DPT processes yield first resolution levels that include between them the resolution levels that would appear out of an equivalent Median Decomposition, since the median operators involved are in the $LULU$-interval. However at later resolution levels this need not be, and some order holds only on average, since the sequence that is passed on from the mth separator to the $m+1$th separator is between those of the other two.

The theorems above therefore not only justify the corresponding DPT procedures, but also, partly, the related Median decomposition, which has until now been heuristic in the motivation. It has, in fact almost none of the strong consistency involved in the $LULU$-cases, but in practice is "almost everywhere" consistent. Concensus is that it "works well". Clearly it can be expected to work well if both the $LULU$-decompositions work well, and give similar decompositions.

In the case of the two $LULU$-decompositions there is a further consistency, that is relatively easy to prove, and is easily motivated. In the case of a discrete Fourier Transform each "resolution level" (frequency level) is a two-dimensional space of sequences.

Since the mapping from a sequence x onto $\alpha_n^- \sin(ni) + \alpha_n^+ \cos(ni)$ is simply a transformation to a different basis of which two vectors are $\sin(ni)$ and $\cos(ni)$, it is clear that the sequence of arbitrary linear combinations of the sin and cos components is decomposed consistently, so that each component comes out with its corresponding amplitude.

The pulse decomposition decomposes a sequence x onto different resolution levels. In each of these, $D_n(x)$ consists of positive and negative pulses, namely

$D_n^-(x) = (I - U_n)C_{n-1}(x)$, which is negative where it is not zero, and

$D_n^+(x) = (I - L_n)U_nC_{n-1}(x)$, which is positive when it is not zero.

This is because $I - U_n \leq 0$ and $(I - L_n) \geq 0$.

Clearly $(I-U_n)C_{n-1}(x)+(I-L_n)U_nC_{n-1}(x) = (I-L_nU_n)C_{n-1}(x) = D_n(x)$, and the crucial interest is whether a further consistency is present, in that, for all $\alpha_j^-, \alpha_j^+ \geq 0$,

$$z = \sum_{i=1}^{n} \alpha_i^- D_i^-(x) + \alpha_i^+ D_i^+(x)$$

is decomposed consistently to yield $D_n^-(z) = \alpha_n^- D_n^-(x)$ and $D_n^+(z) = \alpha_n^+ D_n^+(x)$. This should hold for each $n > 0$.

The next theorem proves this, and has further useful consequences, but needs the following lemma. Defining, as usual,

$$(B_+x)_i = \begin{cases} Bx_i & \text{if } Bx_i > 0 \\ 0 & \text{otherwise,} \end{cases} \quad \text{and} \quad (B_-x)_i = \begin{cases} Bx_i & \text{if } Bx_i < 0 \\ 0 & \text{otherwise.} \end{cases}$$

Lemma.

$$((I - L_nU_n)C_{n-1})_- = (I - U_n)C_{n-1} = D_n^-,$$
$$((I - L_nU_n)C_{n-1})_+ = (I - L_n)U_nC_{n-1} = D_n^+.$$

Proof. $(I-U_n)C_{n-1} \leq 0$, (the zero operator) since $I-U_n \leq 0$ and $(I-L_n)U_nC_{n-1} \geq 0$, since $I - L_n \geq 0$. What is needed to complete the proof is to show that the set of indexes i, where $D_n^- x_i < 0$ and the set of indexes where $D_n^+ x_i > 0$ are disjoint.

Let x be a sequence. Then $C_{n-1}x = z$ is in $\mathcal{M}_{n-1}$. Assume that $D_n^- z_i < 0$, and $D_n^- z_{i-1} = z_{i-1}$. Then

$$(U_n z)_i = \min\{\max\{z_{i-n}, \ldots, z_i\}, \ldots, \max\{z_i, \ldots, z_{i+n}\}\} > z_i.$$

To the left of z_i and to the right of z_i there are $z_\ell, z_r > z_i$ with $|r - \ell| < n+1$. Since $z \in \mathcal{M}_{n-1}$ we have a constant section $z_i = z_{i+1} = \cdots = z_{i+n-1}$, with $U_n z_i = \min\{z_{i-1}, z_{i+n}\} > z_i$.

This is because $z_\ell > z_i$ are both in $\{z_{i-n}, \ldots, z_i\}$ which is therefore monotone decreasing as $z \in \mathcal{M}_{n-1}$. Similarly $\{z_i, \ldots, z_r, \ldots, z_{i+n}\}$ is monotone increasing, and the intersection must be constant. But if $j \in [i, i+n]$, then $(L_n U_n z)_j = \max\{\min\{U_n z_{i-n}, \ldots, U_n z_i\}, \ldots, \min\{U_n z_i, \ldots, U_n z_{i+n}\}\}$ has at least one of the minima equal to $U_n z_i = U_n z_j$, therefore $L_n U_n z_j = U_n z$, and $D_n^+ z_j = 0$. Thus $D_n^+ z$ is zero where $D_n^- z$ is negative. Clearly therefore $D_n^+ z$ can only be positive where $D_n^- z$ is zero.

The other part of the theorem is proved similarly, or by duality. □

Since $(I - U_n)C_{n-1}$ and $(I - L_n)U_n C_{n-1}$ are ftp, the following theorem is easy to prove in analogy with the previous one.

Theorem 8.5. Let $x \in \mathcal{M}_0$ and $DPT(x) = [D_1 x, D_2 x, \ldots, D_N x, D_0 x]$, with $D_0 x = C_N(x)$.

$$z = \sum_{i=1}^{N} (\alpha_i^- D_n^-(x) + \alpha_i^+ D_n^+(x))$$

is decomposed consistently.

Proof. The proof is done as in Theorem 7, with $\alpha_i D_i x$ replaced by $\alpha_i^- D_i^- x + \alpha_i^+ D_i^+ x$.

$$C_j z = L_j U_j(\alpha_j^- D_j^- + \alpha_j^+ D_j^+ + \sum_{i=j+1}^{N} (\alpha_i^- D_i^- + \alpha_i^+ D_i^+))C_{j-1}(x).$$

Applying U_j first we get

$$\begin{aligned} U_j z &= U_j \left(\alpha_j^- D_j^- + \left(\alpha_j^+ D_j^+ + \sum_{i=j+1}^{N} (\alpha_i^- D_i^- + \alpha_i^+ D_i^+) \right) \right) C_{j-1}(x) \\ &= U_j(\alpha_j^-(I - U_j) + (A_{j+1} L_j)U_j)C_{j-1}(x), \end{aligned}$$

with A_{j+1} ftp. This implies by Theorem 4, (since $C_{j-1}(x) \in \mathcal{M}_{j-1}$), that

$$\begin{aligned} U_j z &= (\alpha_j^- U_j(I - U_j) + A_{j+1} L_j U_j)C_{j-1}(x) \\ &= A_{j+1} L_j U_j C_{j-1}(x), \text{ since } U_j(I - U_j) = 0. \end{aligned}$$

Applying L_j to $U_j z$ yields in a similar fashion that

$$C_j z = \sum_{i=j+1}^{n} (\alpha_i^- D_i^- + \alpha_i^+ D_i^+) C_{j-1}(x).$$

The resolution components that are removed at resolution layer j are therefore $\alpha_j^- D_j^- C_{j-1}(x) = \alpha_j^- D_j^-(x)$ and $\alpha_j^+ Dj^+(x)$ and the induction argument used in Theorem 6 completes the proof. □

In a wavelet decomposition, which is just a basis transformation, it is clear that if each individual wavelet in a decomposition of a sequence x is multiplied by a constant and the sum is decomposed, then it decomposes consistently in the sense that the exact multiple of each wavelet appears in the resolution level it came from. Since the pulse decomposition with $LULU$-operators is nonlinear there is no basis transformation, yet the remarkable result of the previous theorem provokes an almost unthinkable idea: That if each individual pulse in a resolution level is multiplied by its own non-negative constant, the sum could still be decomposed consistently!

If this idea is correct the implication for image processing would be very important. For example, if a square white slab were barely visible against a background that is nearly as white, a pulse decomposition on the rows (or columns) of a matrix of luminosity values obtained from a photo can locate the slab by its expected ratio of width to height. Highlighting this particular pulse would make the slab clearly visible, without distorting any of the surroundings. Quite considerable testing has confirmed the idea, but an elegant or illuminating proof seems elusive. Proofs in the case of only a few resolution levels exist, but do not provide sufficient insight to complete even with laborious calculations. Promising ideas are being pursued, which have the merit of classifying many different useful ideas that are not addressed here. We therefore state the idea as a conjecture:

The Highlighting Conjecture. Let z be a sequence formed by adding all the resolution levels of x, after multiplying each particular pulse at a resolution level by a chosen non-negative number. Then z decomposes consistently.

In the following example a random sequence x is generated and $(I - U_1)x$ and $(I - L_1)U_1 x$, in Figure 8.4 below, have each of their pulses amplified by a positive random number. All the resolution levels are then added together again and the sum is decomposed. Comparing the first resolution levels demonstrates the identical modified sequences.

The last few theorems that have led to the establishment of a guaranteed type of consistency in the Discrete Pulse Transform (DPT) are perhaps stronger than the results up to here indicate. Noting that many related operators that are used in various decompositions, for the purpose of fast economical coding and transmission of data, often have sufficiently many of the properties like idempotence, co-idempotence, syntoneness, or commuting with the LULU-operators, to permit proof of further levels of consistency of the DPT. Some typical examples

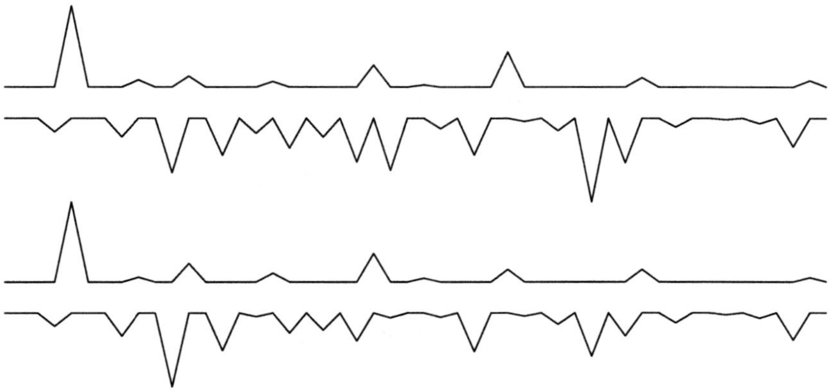

Figure 8.4. The sequences $(I - U_1)x$ and $(I - L_1)U_1x$ and their modified version.

of such operators can be given with some of the relevant properties. Firstly, the following lemma is clear and trivial to prove.

Lemma. If an operator P on $\mathcal{M}_0$ has a support of 1, (that is $(Px)_i = f(x_i)$, where $f(x)$ is a non-negative, non-decreasing, real-valued function, then P is syntone; and neighbor trend preserving (ntp).

The following operators are examples that might occur in practice, and can be demonstrated to have simple properties.

Definition. The Rounding Operator R. For $x \in \mathcal{M}_0$ and index i;

$$(Rx)_i \equiv h\,\mathrm{sgn}(x_i)\mathrm{int}\left(\frac{|x_i|}{h} + \frac{1}{2}\right).$$

Properties.

(a) R is idempotent, syntone, ntp and $R(-x) = -R(x)$.

(b) R commutes with $\vee$ and $\wedge$: $R\vee = \vee R$ and $R\wedge = \wedge R$.

(c) R commutes with all compositions of L_n and U_n.

Proof. (a)

$$\begin{aligned} R(Rx)_i &= h\,\mathrm{sgn}(Rx_i)\mathrm{int}\left(\frac{|Rx_i|}{h} + \frac{1}{2}\right) \\ &= h\,\mathrm{sgn}(x_i)\mathrm{int}\left(\frac{|h\mathrm{sgn}(x_i)\mathrm{int}(\frac{|x_i|}{h} + \frac{1}{2})|}{h} + \frac{1}{2}\right) \\ &= h\,\mathrm{sgn}(x_i)\mathrm{int}\left(\mathrm{int}\left(\frac{|x_i|}{h} + \frac{1}{2}\right) + \frac{1}{2}\right) = Rx_i. \end{aligned}$$

R is syntone by the previous lemma and

$$\begin{aligned}(R(-x))_i &= h\,\mathrm{sgn}(-x_i)\mathrm{int}\left(\frac{|-x_i|}{h}+\frac{1}{2}\right)\\ &= -h\,\mathrm{sgn}(x_i)\mathrm{int}\left(\frac{|x_i|}{h}+\frac{1}{2}\right) = -(Rx)_i.\end{aligned}$$

(b) Consider

$$(R\vee)x_i = h\,\mathrm{sgn}(\vee x_i)\mathrm{int}\left(\frac{|\vee x_i|}{h}+\frac{1}{2}\right).$$

If $x_{i+1} \geq x_i \geq 0$ (similarly if $x_i \geq x_{i+1} \geq 0$) then;

$$\begin{aligned}(R\vee)x_i &= h\,\mathrm{sgn}(x_{i+1})\mathrm{int}\left(\frac{|x_{i+1}|}{h}+\frac{1}{2}\right)\\ &= \max\{Rx_{i+1}, Rx_i\} = (\vee R)x_i.\end{aligned}$$

If $x_{i+1} \geq 0 > x_i$ (similarly if $x_i \geq 0 > x_{i+1}$) then;

$$\begin{aligned}(R\vee)x_i &= h\,\mathrm{sgn}(x_{i+1})\mathrm{int}\left(\frac{|x_{i+1}|}{h}+\frac{1}{2}\right)\\ &= \max\left\{h\,\mathrm{sgn}(x_{i+1})\mathrm{int}\left(\frac{|x_{i+1}|}{h}+\frac{1}{2}\right), h\mathrm{sgn}(x_i)\mathrm{int}\left(\frac{|x_i|}{h}+\frac{1}{2}\right)\right\}\\ &= (\vee R)x_i.\end{aligned}$$

If $0 \geq x_{i+1} > x_i$ (similarly if $0 \geq x_i > x_{i+1}$) then;

$$\begin{aligned}(R\vee)x_i &= h\,\mathrm{sgn}(\vee x_i)\mathrm{int}\left(\frac{|\vee x_i|}{h}+\frac{1}{2}\right) = -h\mathrm{int}\left(\frac{|x_{i+1}|}{h}+\frac{1}{2}\right)\\ &= \max\left\{-h\,\mathrm{int}\left(\frac{|x_{i+1}|}{h}+\frac{1}{2}\right), -h\mathrm{int}\left(\frac{|x_i|}{h}+\frac{1}{2}\right)\right\}\\ &= (\vee R)x_i, \text{ since } |x_i| > |x_{i+1}|.\end{aligned}$$

In all cases therefore $(R\vee)x_i = (\vee R)x_i$.

(c) Clearly $R\vee^n = \vee^n R$ by induction, and similarly $R\wedge^n = \wedge^n R$. Similarly

$$\begin{aligned}RL_n &= R(\vee^n\wedge^n) = (T\vee^n)\wedge^n = (\vee^n T)\wedge^n\\ &= \vee^n(R\wedge^n)\,\vee^n\,(\wedge^n R) = (\vee^n\wedge^n)R = L_n R.\end{aligned}$$

For all compositions of L_n and U_n a similar argument proves commutativity with R. □

Note: R is "nearly" co-idempotent. If we were to replace the "upward rounding" by "downward rounding" to the nearest multiple of h and noting that then $\frac{-h}{2} < x_i - Rx_i < \frac{h}{2}$, it is clear that

$$\frac{|x_i - Rx_i|}{h} < \frac{1}{2} \quad \text{so that} \quad \text{int}\left(\frac{|x_i - Rx_i|}{h} + \frac{1}{2}\right) = 0.$$

Hence $R(I - R) = 0$.

(This can be achieved by subtracting the smallest machine number eps from the argument of the integer function, or a similar idea.)

A similar operator is a truncating operator of the type

$$(Tx)_i = h \operatorname{sgn}(x_i)\text{int}\left(\frac{|x_i|}{h}\right),$$

where we expect similar properties, with the addition of co-idempotence.

Definition. A Thresholding Operator T. $\forall x \in \mathcal{M}_0$,

$$(T_t x)_i = (Tx)_i \equiv \begin{cases} x_i & \text{if } |x_i| \le t, \\ \operatorname{sgn}(x_i)t & \text{if } |x_i| \ge t \end{cases}.$$

Properties.
(a) T is Idempotent and $T(-x) = -Tx$.
(b) T is fully trend preserving (ftp).
(c) T commutes with $\vee$ and $\wedge$: $T\vee = \vee T$ and $T\wedge = \wedge T$.
(d) T commutes with $L_n, U_n, L_nU_n, U_nL_n, C_n$ and F_n.

Proof. (a) Trivial, or by the previous lemma.
(b) Let $x_{i+1} \ge x_i$. ($x_i \ge x_{i+1}$ is handled similarly). Then $(Tx)_{i+1} \ge (Tx)_i$ trivially, so that T is ntp. Consider $(I - T)x_i = x_i - (Tx)_i$ and $(I - T)x_{i+1} = x_{i+1} - (Tx)_{i+1}$. If $x_i, x_{i+1} \in [-t, t]$, then both terms are zero. If $x_i \in [-t, t]$ but $x_{i+1} > t$, the first term is zero and the second is positive. If $x_{i+1} \in [-t, t]$ but $x_i < -t$, then the first term is negative and the second zero. If $x_i, x_{i+1} \notin [-t, t]$, then $x_{i+1} > t > -t > x_i$. Then the first is negative and the second positive. In each case therefore $(I - T)x_{i+1} \ge (I - T)x_i$, and $I - T$ is ntp.
(c) Consider $(\vee T)x_i = \max\{Tx_i, Tx_{i+1}\}$. If $x_{i+1}, x_i \in [-t, t]$, then $(\vee T)x_i = \vee x_i = \max\{x_i, x_{i+1}\} = (T\vee)x_i$. If $x_{i+1} \in [-t, t]$ and
(a) $x_i > t$, then $(\vee T)x_i = \max\{t, x_{i+1}\} = t = (T\vee)x_i$,
(b) $x_i < -t$, then $(\vee T)x_i = \max\{x_{i+1}, -t\} = x_{i+1} = (T\vee)x_i$.
(Similarly if $x_i \in [-t, t]$ and $x_{i+1} \notin [-t, t]$.)

If $x_{i+1}, x_i \notin [-t, t]$ and
(a) $x_{i+1}, x_i > t$; then $(\vee T)x_i = \max\{t, t\} = (T\vee)x_i$,
(b) $x_{i+1}, x + i < -t$, then $(\vee T)x_i = \max\{-t, t\} = (T\vee)x_i$,
(c) $x_{i+1} > t > -t > x_i$, then $(\vee T)x_i = \max\{-t, t\} = t = (T\vee)x_i$,
(d) $x_i > t > -t > x_{i+1}$, then $(\vee T)x_i = \max\{t, -t\} = t = (T\vee)x_i$.
Therefore T commutes with $\vee$, and a similar argument proves that $T\wedge = \wedge T$.

(d) Clearly $T\vee^n = \vee^n T$ by induction and similarly $T\wedge^n = \wedge^n T$. Similarly

$$\begin{aligned} TL_n &= T(\vee^n\wedge^n) = (T\vee^n)\wedge^n = (\vee^n T)\wedge^n \\ &= \vee^n(T\wedge^n) = \vee^n(\wedge^n T) = (\vee^n\wedge^n)T = L_n T. \end{aligned}$$

For all other compositions of L_n and U_n a similar argument proves commutativity. □

Note: T is not co-idempotent, as a simple example of a constant sequence $x_i = c = 3t$ shows. $T(I-T)x = T(c-Tc) = T(3t-t) = x \neq 0$.

A similar operator is another thresholding operator;

$$(\perp x)_i = \begin{cases} x_i & \text{if } |x_i| > t, \\ 0 & \text{if } |x_i| < t. \end{cases}$$

Similar properties to the above case can be expected, as well as co-idempotence.

With the properties of the above operators it seems that the effect on the DDT can be investigated with confidence, when this is required. Clearly if the conjecture following Theorem 8 is true we expect consistent decomposition after modification of the resolution levels by the above operators.

From the definition of $LULU$-operators it appears to have a considerable computational complexity, but for the purpose of successive decompositions by L_nU_n (or U_nL_n) there are considerable computational simplifications. The example with L_nU_n can be chosen to illustrate. Given a sequence $x = \{x_i : i = 1, 2, \ldots, N\}$ in $\mathcal{M}_0$ the first decomposition is with U_1, followed by L_1 to yield a sequence L_1U_1x in $\mathcal{M}_1$ and a residual $R_1x = x - L_1U_1x$. At each stage, the sequence that is decomposed by L_nU_n is in $\mathcal{M}_{n-1}$, which permits a simplified calculation. Consider $(U_nx)_i = \min\{\max\{x_{i-n}, \ldots, x_i\}, \ldots, \max\{x_i, \ldots, x_{i+n}\}$. Near endpoints, where left (or right-) neighbors are not defined, it is simple to omit them in the maximum (or minimum) calculations. This is clearly equivalent to appending sufficient values equal to x_1 to the left (and equal to x_N at the right). Given that $n \geq 1$ and x is $(n-1)$-monotone, the first (and the last) values of U_nx can be copied from x. This is because the following arguments hold. Since $\{x_k, \ldots, x_{k+n+1}\}$ is monotone;

(i) Assume $x_1 \leq x_2 \leq \cdots \leq x_{n+1}$, then for $j = 1, \ldots, n$,

$$\begin{aligned} U_nx_j = \min\{&\max\{x_1, \ldots, x_j\}, \max\{x_2, \ldots, x_j, x_{j+1}\}, \ldots \\ &\ldots, \max\{x_j, \ldots, x_j\} = x_j, \end{aligned}$$

since $x_j = \max\{x_1, \ldots, x_j\}$ and x_j is not larger than all the others since they are upper bounds.

(ii) Assume $x_1 \geq x_2 \geq \cdots \geq x_{n+1}$. Noting that $x_n \geq x_{n+1}$ implies that $x_n \geq x_{n+1} \geq \cdots \geq x_{n+n}$, it is clear that $\max\{x_j, x_{j+1}, \ldots, x_{j+n}\} = x_j$, so that

$$U_n x_j = \min\{\max\{1, \ldots, x_j\}, \ldots, \max\{x_j, x_{j+1}, \ldots, x_{j+n}\} = x_j,$$

since the last maximum is x_j and the previous cannot be less.

Thus $U_n x_j = x_j$ for $j \in \{1, \ldots, n\}$.

A similar argument also holds for $j \in \{N - n + 1, \ldots, N\}$ so that $2n$ values of $U_n x_j$ are already calculated.

The following sequences are calculated successively.

$$\begin{aligned} e_i &= \max\{x_i, x_{i+1}, \ldots, x_{i+n}\} = \max\{x_i, x_{i+n}\}, & i &= 2 \text{ to } N - n - 1, \\ t_i &= \min\{e_{i-n}, \ldots, e_i\} = \min\{e_{i-n}, e_i\}, & \text{for } i &= n + 1 \text{ to } N - n - 1. \end{aligned}$$

This requires a total of $2N - 3n - 4$ comparisons and yields $U_n x$.

This is followed by a similar process for $L_n U_n x$.

For a full decomposition up to and including level n therefore, a total of

$$n(4N - 3n - 7) = \sum_{i=1}^{N} (4N - 6i - 4)$$

comparisons are needed. To compute the residual $x - L_n U_n x$ at each level, only values different from 0 need a subtraction. (The minority of points, definitely less than $\frac{N}{n+1}$, but generally much less.) There is no point in letting n be more than $N - 1$, since then $x - L_n U_n x = 0$ and $L_n U_n x$ is a constant. Generally therefore the total number of comparisons is less than $N^2 - 11N + 8$. There are indications that a further saving can be achieved, as the above estimates are not sharp. Certainly it is clear that a great amount of parallelization is possible, depending upon the equipment available. Some simple examples demonstrate very economic coding if some structures are present. Such economizing is likely to occur often when quantizing (and/or thresholding) is introduced for the amplitude, as is the case with measurements in fixed bit format It should be noted that in such a case none but a finite number of possible numbers available are sufficient for all decompositions, since the only operations performed are selections and differences. This is an important economization, compared to other values that are introduced by wavelet decompositions, where averages and differences are required.

There are a variety of alternative computational procedures that can be researched for further economization, depending on the number of resolution levels required. This vast field of algorithmic analysis is still open.

Example 1. Let $N = 2k$ and $x = \{x_i = |i - k|\}$.

At level 1.	$(x - L_1U_1x)_i$	$= \delta_{ik}$ (The Kronecker delta)
At level 3.	$(x - L_3U_3)L_2U_2L_1U_1x_i$	$= \delta_{ik-1}$
At level 5.		$= \delta_{kk-2}$
$\vdots$		$\vdots$
At level $2m + 1$.		$= \delta_{ik-m}$ up to $2N - 1$.

Thus there are precisely N pulses. Each is of width $2m-1$ and constant amplitude, and coding requires N starting indexes and N amplitudes.

Example 2. Let $N = 2k$ and $x = \{x_i = (-1)^i\}$. Then, at level 1, there are pulses at each even integer, except 1, and at each odd integer, after which there is only one pulse of width $N - 1$ and a remaining value of -1 (or a pulse of width N and amplitude -1). Thus a total of k indexes and $k + 2$ amplitudes.

Example 3. When random noise from a cubic-B-spline distribution is added to three uniform level pulses of duration 1, 2 and 3, the visually significant counts are $23, 6$ and 5 out of a sequence length of 50. Quantizing and thresholding appropriately, and reconstituting from only three significant pulses would yield an image that is substantially the significant features from three indexes and amplitudes. The thresholds can be chosen a priori to estimates corresponding to amplitudes and distribution of the random noise. Impulsive noise of a prescribed width can be removed by omitting higher levels altogether.

One important aspect of the decomposition of an arbitrary sequence, the question of the total number of pulses in the resolution levels, can be answered by the following simple argument:

Taking the case of a finite sequence of N points and appending zeroes (or assuming periodicity with $x_{N+i} - x_i$) on either side, it becomes clear that a pulse appears in a resolution level when one of two neighboring values of the sequence smoothed by some L_nU_n (or U_nL_n) is replaced by the other. A pulse therefore has the amplitude of such a difference. The total variation contains not more than $N + 1$ such differences and each pulse removes at least one such term (Theorem 7.1). The last pulse to be removed takes with it two such terms. Therefore no more than N pulses appear in total and the original sequence can be fully reconstructed from these. This is verified by the chosen examples above, and the value of the constructive proof of Theorem 7.1 in substantiating this argument is apparent. Apart from the overheads of keeping track of the index at which a particular pulse is located, the coding economy of storing/transmitting the resolution levels is no more than that of the original sequence. Since a sequence is generally contaminated by some random noise from a distribution, we can expect from experience that this appears mainly in the higher resolution levels, where it could be removed with minimal change to the essential features remaining.

How the identically, independently distributed (i.i.d.-) random noise will be distributed by the PPT seems the key question here. The operators are not linear and the addition of such noise to a signal will not yield a decomposition that is predictable if the decompositions of signal and noises are understood separately. Nevertheless, practical experience suggests that the behavior can be expected to be fairly well predicted by understanding the case of pure noise. The issue is being researched and results are promising.

At this stage it is however instructive to introduce the basic initial results and ideas in the context of some fair comparisons of the $LULU$-operators with linear filters of comparable type.

9. Fair Comparison with Linear Smoothers

Seit ich des Suchens Müde ward
Erlernte ich das Finden.
Seit mir ein Wind hielt Widerpart
Segl' ich mit allen Winden.

Nietzsche

A theory of linear (digital-) filters is well known and established. A good introduction is provided by Hamming (11). The underlying ideas can however be briefly, and casually, introduced, for the purpose of comparison with other smoothers, specifically the *LULU*-operators and popular selectors and compositions.

As in the case of smoothers like the (running-) median Mn, a window W_n with width $2n+1$ considers the subset $\{x_{i-n}, \ldots, x_i, \ldots, x_{i+n}\}$ of a sequence x, for successive i. This window can be considered to "move over" the sequence x, or the sequence can be considered to "pass through" the window. The elements in the window are used to compute an output value

$$(F_n x)_i = \sum_{j=i-n}^{i+n} \alpha_j x_j = \sum_{k=-n}^{n} \alpha_k x_{i-k},$$

with α_j constants.

The simplest way of motivating this type of smoother is to consider it as a local least squares projection of the values $x_{i-n}, \ldots, x_{i+n}$ onto a best constant value, and using this value to replace x_i. This corresponds to the case of the median smoother which does the same, but projects in the ℓ_1-norm, so as to be more "robust".

This least squares local projection yields the value

$$(F_n x)_i = \sum_{k=-n}^{n} \frac{1}{2n+1} x_{i+k}.$$

Clearly the operator F_n maps sequences in X onto sequences in X and meets all the requirements of a smoother. The important requirements are easy to verify; F_n is axis independent and scale independent. If also preserves sequences that are of the form $x = \langle x_i = \alpha i + \beta \rangle$, where α and β are constants.

A similar argument, using local projections to replace x_i by the value at i of the least squares quadratic polynomial to the values $\{x_{i-n}, \ldots, x_{i+n}\}$, yields, for example with $n = 2$ the output,

$$y_i = (F_2 x)_i = \frac{1}{35}[-3x_{i-2} + 12x_{i-1} + 17x_i + 12x_{i+1} - 3x_{i+2}].$$

It becomes clear that, in each case, there is a weighted average of the values in the window that replaces x_i with the smoothed value y_i.

For the second example it is easy to verify that it preserves sequences that are sampled quadratic polynomials, which results in a better "shape preservation" of a sequence. The penalty is that the weighted average is not a convex combination anymore, since, although the coefficients $-\frac{3}{13}, \frac{12}{35}, \frac{17}{35}, \frac{12}{35}, -\frac{3}{35}$ add up to 1 (preserving constant sequences), the first and last are negative.

Smoothers of the above type are easily seen to be linear and the set of all such smoothers form a vector subspace of the set of smoothers from X to X. Such a smoother can be considered to yield an output that is, for each $x \in X$, a convolute of x with a specific constant sequence α.

Definition. A sequence y is the (discrete) convolute of sequences x and α

$$\text{iff. } y_i = \sum_{j=-\infty}^{\infty} x_j \alpha_{i-j}.$$

Smoothers (filters) of the following types appear in literature under the names "FIR filters" ("finite response filters"), "transversal filters", "tapped delay line filters", "moving Average filters" or just "non-recursive (digital-) filters". They shall be called non-recursive filters here, or just filters, where there is no misunderstanding likely.

It is often useful to use values that are already smoothed in the window, so that the output is of the form

$$y_i = \sum_{i=-\infty}^{\infty} \alpha_j x_{i-j} + \sum_{j=-\infty}^{\infty} \beta_j y_{i-j}, \quad \text{with } \alpha, \beta \text{ constant sequences.}$$

This is an implicit definition of y, so that evaluation is awkward, and generally the version that is recursively calculable is used, where β_j is chosen 0 for j non-positive.

Such smoothers are called "infinite impulse response filters", "ladder filters", "lattice filters", "autoregressive moving average filters", "ARIMA-filters". Here they shall be called recursive filters.

Well-known versions are:

a) $y_i = y_{i-1} + \frac{1}{2}(x_i + x_{i-1})$. (The Trapezium Rule for integration.)
b) $y_i = \alpha x_i + (1-\alpha) y_{i-1}$, $\alpha \in (0,1)$. (The exponential smoothing prediction.)
c) $y_i = \alpha(x_i - x_{i-1}) + (1-\alpha) y_{i-1}, \alpha \in (0,1)$. (The "trend indicator".)

Noting that sequences formed by sampling of polynomials are not generally absolutely summable, the above ideas of local projection could be generalized to rather yield local projections (or global projections onto sampled spline functions). This idea is inherent in the currently very active research field of wavelet analysis, specifically spline wavelet analysis, where a sequence is mapped onto a spline function. The spline function undergoes a wavelet decomposition, and smoothing simply is the removal of "high-frequency" wavelets. The idea for semi-orthogonal spline wavelets is an illuminating version of general spline smoothing as done by Schoenberg (28). Wavelet smoothers are not translation invariant and therefore, as previously argued, have additional complications to attend to.

For the purpose of comparison the usual digital filters will however provisionally suffice. They are traditionally popular because of the electronic analogy in RC-networks, and the smoothing (filtering) has therefore developed in conjunction with electronic transmission of signals where sinusoidal electric vibrations are used as carriers with other (locally-) periodic information superimposed.

The sequences $x = \cos(vi)$ and $y = \sin(vi)$, for various frequencies v, are thus naturally prevalent. Since these are all eigensequences of linear mappings from X to X, the natural ideas used for analysis are Fourier Analysis and projections onto subspaces of sequences such as x and y.

The idea of a frequency is central, and will be measured in cycles per second (Herz) or in radians per second (angular frequency). Sequences of constant frequency are periodic, and the fundamental period is the shortest interval within which the sequence repeats itself.

Definition. Let T_N be the subspace of (periodic) functions of the form

$$f(x) = \alpha_0 + \sum_{i=1}^{N} \alpha_i \cos(ix) + \beta_i \sin(ix)$$

and F_N be the subspace of (periodic) sequences that are samplings of these at a given sampling rate.

With the above definition of the signals (sequences in F_N) it is soon clear that the requirement that a smoother preserves sequences in F_N is difficult to meet as there is a fundamental ambiguity related to the concept of "aliasing". Aliasing is observed in typical "Western" films, where the spokes of a coach which are observed to undergo a peculiar behavior when the coach accelerating. Stroboscopic light flashes of a sinusoidal movement produces similar effects. Aliasing is simply due to the fact that frequencies higher than sampling frequency ("Nyquist-frequency") are undistinguishable from a lower frequency.

Consider the sequence $x(i) = \cos(2\pi(n+\alpha)i + \beta)$, where n is an integer, α a positive fraction and β a fixed (phase-) angle.

$$x_i = x(i) = \cos(2\pi(n+\alpha)i + \beta) = \cos(2\pi\alpha i + \beta).$$

It is clear that every cosine sequence is thus indistinguishable from a particular cosine frequency with angular frequency between 0 and $\frac{1}{2}$.

High frequencies are thus aliased under a low frequency, and only sequences with frequencies between 0 and $\frac{1}{2}$ can be preserved by a filter using the sampled values $x = \langle x_i \rangle$. When a general sequence that is periodic in a section is filtered, the result has a fundamental ambiguity as to the original frequency sampled. This is an analogy of the fundamental ambiguity for impulsive noise, which yields the $LULU$-interval. Noting that a filter yields a unique sequence in some subspace F_N, it is clear that it does not warn us of the ambiguity present in the smoothed sequence. Even when a slowly increasing frequency is sampled, the approach of the first multiple of the Nyquist frequency is difficult to detect, whereas the separation of the two smoothed samplings LUx and ULx explicitly warns of ambiguity.

It is also noteworthy to remark that aliasing is not restricted to trigonometric polynomials but to all polynomials, and other classes of functions as well. This problem of aliasing should be kept in mind during all the subsequent analysis of filters and smoothers in general, and viewed as a fundamental ambiguity as to the signal content in a sequence.

For linear filters, as with linear operators in general, the general idea for analysis is eigenanalysis.

Consider a (linear-) filter Ln mapping $x \in X$ onto $y \in X$ such that

$$y_i = (Lnx)_i = \sum_{j=-n}^{n} \delta_j x_{i-j}.$$

Let $\quad x_i = \sin(\alpha i + \theta)$. $\quad$ Then elementary trigonometric identities yield

$$y_i = (Lx)_i = \sum_{j=-n}^{n} \delta_j \sin(\alpha(i+j)+\theta) = \sum_{j=-n}^{n} \delta_j (\sin(\alpha i + \theta + \alpha_j)$$
$$= A \sin(\alpha i + \theta) + B\cos(\alpha i + \theta) = \sqrt{A^2+B^2}\sin(\alpha i + \theta + \beta),$$
$$\text{with} \quad A = \sum_{j=-n}^{n} \delta_j \cos\alpha j, \quad B = \sum_{j=-n}^{n} \delta_j \sin\alpha j \text{ en } \beta = \arctan(\frac{B}{A}).$$

The sequence thus undergoes a phase change of β and an amplitude change of $\sqrt{A^2+B^2}$. When a filter is symmetric ($\delta_j = \delta_{-j}$) it is clear that

$$B = \sum_{j=-n}^{n} (\delta_j - \delta_{-j}) \sin\alpha_j = 0,$$

so that the phase change is zero, and

$$(Lx)_i = A\sin(\alpha i + \theta).$$

Thus the sequence x_i is an eigensequence with eigenvalue A.

When $\sum_{j=-n}^{n} \delta_j = 1$, which is required for the preservation of constant sequences by smoothers,

$$A = \delta_0 + \sum_{j=1}^{n}(\delta_j + \delta_{-j}) \cos \alpha j = \delta_0 + \sum_{j=1}^{n}(\delta_j + \delta_{-j})(1 - 2\sin^2(\frac{\alpha j}{2}))$$
$$= \sum_{j=-n}^{n} \delta_j - \sum_{j=1}^{n}(\delta_j + \delta_{-j}) 2\sin^2(\frac{\alpha}{2}j).$$

The first term is 1 and the second is 0 at $\alpha = 0$, and grows as α grows $\delta_j \geq 0$. Thus the low frequencies are almost unchanged but higher frequencies are decreased in amplitude by the filter.

For the specific case of the filter with coefficients $[\frac{1}{5}, \frac{1}{5}, \frac{1}{5}, \frac{1}{5}, \frac{1}{5}]$,

$$A = \frac{1}{5}\sum_{j=-2}^{2} \cos \alpha j = \frac{1}{5}[1 + 2\cos\alpha + 2\cos 2\alpha] = \frac{1}{5}\left(\frac{\sin\frac{5}{2}\alpha}{\sin\frac{\alpha}{2}}\right).$$

Similarly, for a moving average filter with $\delta_j = \dfrac{1}{2n+1}$ for $j \in [-n, n]$ it follows that

$$A(\alpha) = \frac{1}{2n+1} + \sum_{j=1}^{n} \frac{2\cos(\alpha j)}{2n+1} = \frac{\sin(n\alpha + \frac{\alpha}{2})}{(2n+1)\sin\frac{\alpha}{2}},$$

since

$$\sum_{j=1}^{n} 2\cos\alpha j \sin\frac{\alpha}{2} = \sum_{j=1}^{n}[\sin(\alpha j + \frac{\alpha}{2}) - \sin(\alpha j - \frac{\alpha}{2})] = \sin(n\alpha + \frac{\alpha}{2}) - \sin\frac{\alpha}{2}.$$

Since the filter is linear, any periodic function will pass through the filter as if its constituent frequencies pass through separately and recombine, and since

$$x_i = \sum_{k=1}^{n} \alpha_k \sin(w_k i) + \beta_k \cos(w_k i) = \sum_{k=1}^{n} \sqrt{\alpha_k^2 + \beta_k^2}\, \sin(w_k i + \theta_k),$$

it follows that

$$y_i = (Lx)_i = \sum_{k=1}^{n} A(w_k)\sqrt{\alpha_k^2 + p_k^2}\, \sin(w_k i + \theta_k)$$
$$= \sum_{k=1}^{n} a(w_k)[\alpha_k \sin(w_k i) + \beta_k \cos(w_k i)].$$

If $\cos x + i\sin x = e^{ix}$ and $\cos x - i\sin x = e^{-ix}$, the trigonometric identities

$$\sin(x+y) = \sin x\cos y + \cos x\sin y \quad \text{and} \quad \cos(x+y) = \cos x\cos y - \sin x\sin y$$

are replaced by

$$e^{ix}e^{iy} = e^{i(x+y)}.$$

It becomes apparent that the complex power sequences are eigensequences of the linear translation operator

$$(Tu)_j = u_{j+1}, \quad \text{since} \quad e^{iw(t+h)} = e^{iwh}e^{iwt} = \lambda(w)e^{iwt}.$$

This idea also is applicable in the following cases.

i) The non-recursive filters of the form $y_n = \sum_{j=-m}^{m} c_j x_{n-j}$ yield, with substitution of $x_j = e^{iwj}$, the output

$$y_n = e^{iwn}\sum_{j=-m}^{m} c_j e^{-iwj} = \lambda(w)e^{iwn},$$

with $\lambda(w) = \sum_{j=-m}^{m} c_j e^{-iwj}$.

ii) The difference operator $\triangle f(t) = f(t+h) - f(t)$, since

$$\triangle e^{iwt} = e^{iw(t+h)} - e^{iwt} = e^{iwt}(e^{iwh} - 1).$$

iii) The summation operator $\sum f(t)$, since

$$\sum_{k=1}^{n} e^{iw(t+kh)} = e^{iwt}\sum_{k=1}^{n} e^{iwhk} = e^{iwt}\left(e^{iwt}\left(\frac{e^{iwnh}-1}{e^{iwh}-1}\right)\right) = \lambda(w)e^{iwt}.$$

The transfer function of a recursive linear filter of the form

$$A(\alpha) = \delta_0 + \sum_{j=1}^{n}(\delta_j + \delta_{-j})\cos(\alpha j)$$

can be considered a cosine transform of the sequence $\langle\delta_j\rangle$. The output

$$y_i = \sum_{j=-n}^{n} \delta_j x_{i-j} = \sum_{-\infty}^{\infty} \delta_j x_{i-j}$$

is considered as a convolute of $\langle x_i\rangle$ and $\langle\delta_j\rangle$, with $\delta_j = 0$ for $|j| > n$.

For a smoother it is required that

$$\sum_{j=-\infty}^{\infty} \delta_j = 1.$$

The transfer function can now be considered as the Fourier transform $\overline{\delta}$ of the sequence δ. The Fourier transform of the sequence x multiplied by $\overline{\delta}$ is the transform of the convolute of the sequence x with δ, and yields the required smoothed x by inverting.

Filtering is thus simply multiplication in the "frequency domain". This is in analogy to the case of functions, as is used in the development of B-splines (8) and in the development of Wavelet Theory. For the design of smoothers, this permits the idea of designing filters as follows.

Since $A(\alpha) = \delta_0 + 2\delta_1 \cos\alpha + 2\delta_2 \cos 2\alpha + \cdots + 2\delta_n \cos n\alpha$ for a symmetric filter, the ideal transfer can be approximated with a cosine polynomial.

Example. Let $T(\alpha) = \begin{cases} 1 & \text{for } |\alpha| < \beta, \\ 0 & \text{otherwise.} \end{cases}$ This is an ideal filter as it is a projection.

A Fourier expansion gives.

$$T(x) = \gamma_0 + \sum_{i=0}^{\infty} \gamma_i \cos i(x),$$

with

$$\gamma_i = \frac{1}{\pi} \int_0^{\pi} T(t) \cos(it)dt = \frac{\sin i\beta}{\pi i}, \quad \text{and} \quad \gamma_0 = \frac{\beta}{\pi}.$$

The filter coefficients are now $\delta_0 = \gamma_0$ and $\delta_i = \dfrac{\gamma_i}{2}$, but the disadvantage is an infinite window. The coefficients also decay with tempo $0(\frac{1}{i})$, because of the discontinuity of T at $\pm\beta$. Other approximations, say with Lanczos factors, would still have slow convergence, since the fundamental Heisenberg inequality (8) assures a wide window if the transform is to be narrow. Recursive filters could improve things somewhat, but not sufficiently, since the support of a filter needs to be small if good shape preservation of signals is desired.

It turns out that there is an alternative set of eigensequences that are common to linear smoothers and some selectors and compositions of these, and thus permits comparison. The sequences that are formed by sampling exponentials present themselves naturally.

Considering the sequences $\alpha = \langle x_i = \alpha^i \rangle$, it is clear that, when $\alpha > 0$, they are monotone. By the previous theory $LUa = LUa = a$, so that a is an eigensequence of the $LULU$-operators, for each n, and also of the medians and similar smoothers inside the $LULU$-interval.

A linear filter F, with coefficient sequence δ, will have an output given by

$$(Fa)_i = \sum_{j=-n}^{n} \delta_j a_{i+j} = \sum_{j=-n}^{n} \delta_j a^{i+j} = \alpha^i \sum_{j=-n}^{n} \delta_j \alpha^j.$$

This means that a is an eigensequence of F with the eigenvalue

$$\lambda_\alpha = \sum_{j=-n}^{n} \delta_j \alpha^j, \quad \text{with} \quad \lambda_\alpha \geq 1.$$

This eigenvalue can be 1 with at most $2n$ values of α, unless $\alpha = 1$. Thus at best only constant sequences can be preserved in general. Areas in which the signal undergoes fast changes (exponential sections) will therefore undergo distortion. In image processing this becomes clear by the typical "edge-smearing" of linear smoothers. In auditory signals this may not be so noticeable, since the ear detects in the "frequency domain" rather than in the "time-domain", and amplitude is perhaps not picked up with the same resolution as frequency. Since edges can be arbitrarily steep, linear smoothers can yield an arbitrarily large distortion.

For negative values of α, a is still an eigensequence and the eigenvalue is now smaller.

For negative α, the medians still have the sequence as an eigensequence, but with a peculiar behavior of the eigenvalue.

$$\text{median}[\alpha^{i-1}, \alpha^i, \alpha^{i+1}\} = \alpha^i \alpha, \text{ if } |\alpha| < 1.$$

Thus when $\alpha = -1$, the eigenvalue -1 results, as was previously established. This is because a is a sequence that is periodic at precisely the Nyquist frequency. In this respect the median applied twice is better, at least not permitting a negative eigenvalue. Linear combinations like $\frac{1}{2}[M_1 + M_1^2]$ and $\frac{1}{4}[M_1 + 2M_1^2 + M_1^3]$ exhibit eigenvalues closer to zero, thus confirming practical experience.

It is easy to show that a is not an eigensequence of the $LULU$-operators when $\alpha < 0$. They yield outputs that provide an "envelope of uncertainty" as

$$\begin{aligned}(L_1 U_1 a)_i &= \max\{\alpha^i, \min\{\alpha^{i+1}, \alpha^{i-1}\}\} \quad \text{and} \\ (U_1 L_1 a)_i &= \min\{\alpha^i, \max\{\alpha^{i+1}, \alpha^{i-1}\}\},\end{aligned}$$

in a growing or diminishing oscillation in a sequence, thus warning of ambiguity. This corresponds to previous argument and experience.

The operator $G = \frac{1}{2}[LU + UL]$ does however have a as eigensequence for $\alpha < 0$.

$$(Ga)_i = \frac{1}{2}(\alpha^i + \alpha^{i+1}) = \alpha^i \left(\frac{1+\alpha}{2}\right).$$

Therefore $\lambda_a = \dfrac{1+\alpha}{2}$ is the eigenvalue corresponding to a.

Similarly

$$(G^n a)_i = \alpha^i \left(\frac{1+\alpha}{2}\right)^n.$$

Thus repeating G results in an output converging to the zero line for $-1 \leq \alpha < 0$, and near $\alpha = -1$ this convergence is rapid.

When comparing smoothers that are index-symmetric the sequences $a = \langle \alpha^i \rangle$ and $\langle (\frac{1}{\alpha})^i \rangle$ are clearly handled identically and thus this eigenanalysis can be restricted to $|\alpha| \leq 1$.

The comparison with this eigenanalysis does not reflect favorably on the linear smoothers. The linear smoothers demonstrate eigenvalues that approximate 1 and 0 only at $\alpha = \pm 1$, whereas medians and some combinations have eigenvalue 1 for all $\alpha > 0$ and approximating 0 better for $\alpha < 0$.

Having proved that LU and UL have only two eigenvalues 1 and 0, an ideal situation, it is clear that their resistance to identifying an oscillation as noise can be considered a quality of merit rather than a defect. On the first measure of comparison therefore, namely eigenanalysis on a common set of eigenvalues, the popular nonlinear smoothers indicate a marked superiority over linear filters. Lastly, the following argument yields reassurance of no amplification of either "signal" or noise for a large class of smoothers.

Clearly $|Sx|_\infty \leq |x|_\infty$ for a selector, so that $|Sx|_\infty = |\lambda x|_\infty = |\lambda|\,|x|_\infty \leq |x|_\infty$. Also if S_ℓ and S_u are selectors and $S_\ell \leq A \leq S_u$, then $S_\ell x \leq Ax \leq S_u x$ for each $x \in X$.

Since $-|x|_\infty \leq (S_\ell x)_i \leq (Ax)_i \leq (S_u x)_i \leq ||x||_\infty$, for each sequence $x \in X$ and each index i, it follows that $||Ax||_\infty \leq ||x||_\infty$.

Another method of comparison, on a level playing field again, is the behavior of random noise. This has received attention in the case of median smoothers (2), but can be considered again, with the inclusion of comparison with $LULU$-smoothers.

We consider firstly a sequence $x = \langle x_i \rangle$ of identically independent random numbers from a probability density (pdf) $f_x(t)$ with

$$F_x(t) = \int_{-a}^{t} f_x(u)du.$$

We consider an operator R_k that selects from the windowed subset

$$W_i = \{x_{i-n}, \ldots, x_{i+n}\}$$

the value ranked kth. Let $(R_k x)_i$ be this value and the operator be defined accordingly. The following theorem is easy to prove and well known, but included for comparison.

Theorem 9.1. The probability that the kth rank is larger than z is given by

$$F_y(z) = P(y_i \leq z) = \sum_{j=k}^{2n+1} \binom{2n+1}{j} F_x^j(z)(1 - F_x(z))^{2n+1-j}.$$

Proof. The probability is given by the sum from k to $2n+1$ of the probabilities that j of the elements of W_i are not larger than z, which is

$$\binom{2n+1}{j} F_x^j(z)(1 - F_x(z))^{2n+1-j},$$

since the elements of W_i can be arranged in $\binom{2n+1}{j}$ ways such that j elements are not larger than z (probability $F_x^j(z)$) and $2n+1-j$ elements are not smaller (probability $(1-F_x(z))^{2n+1-j}$. □

Theorem 9.2. The probability density function of $R_k x_i$ is given by

$$f_y(z) = (2n+1)\binom{2n}{k-1}[F_x^{k-1}(z)(1-F_x(z))^{2n+1-k}]f_x(z).$$

Proof. $F_y(z) = \sum_{j=k}^{2n+1} \binom{2n+1}{j} F_x^j(z)(1-F_x(z))^{2n+1-j}$,

$$\begin{aligned} f_y(z) &= \frac{d}{dz}F_y(z) = f_x(z)\sum_{j=k}^{2n+1}\Bigg[\binom{2n+1}{j} jF_x^{j-1}(z)(1-F_x(z))^{2n-(j-1)} \\ &\quad - \sum_{j=k}^{2n+1}\binom{2n+1}{j}F_x^j(z)(1-F_x(z))^{2n-j}(2n-(j-1))\Bigg] \\ &= f_x(z)\Bigg[T_k + \sum_{j=k+1}^{2n+1}\binom{2n+1}{j} jF_x^{j-1}(z)(1-F_x(z))^{2n-(j-1)} \\ &\quad - \sum_{i=k+1}^{2n+1}\binom{2n+1}{i-1}(2n+1-(i-1))F_x^{i-1}(z)(1-F_x(z))^{2n-(i-1)}\Bigg] \\ &= f_x(z)T_k, \text{ where } T_k = \binom{2n+1}{k} kF_x^{k-1}(z)(1-F_x(z))^{2n-(k-1)}. \end{aligned}$$ □

Examples.

1. The maximum operator R_{2n+1}. If $k = 2n+1$, then

$$F_y(z) = F_x^{2n+1}(z) \quad \text{and} \quad f_y(z) = (2n+1)F_x^{2n}(z)f_x(z).$$

2. The minimum operator R_1.

$$F_y(z) = \sum_{j=1}^{2n+1}\binom{2n+1}{j}F_x^j(z)(1-F_x(z))^{2n+1-j}.$$

But, from the binomial theorem, the sum from 0 to $2n+1$ equals $(F_x(z) + 1 - F_x(z))^{2n+1} = 1$. Thus

$$F_y(z) = 1-(1-F_x(z))^{2n+1} \quad \text{of} \quad 1-F_y(z) = (1-F_x(z))^{2n+1}.$$

3. With $n = 1$ the cumulative distribution of the output of M_1 is given by

$$\begin{aligned} F_y(z) &= \sum_{j=2}^{3}\tbinom{3}{j}F_x^j(z)(1-F_x(z))^{3-j} = 3F_x^2(z)(1-F_x(z)) + F_x^3(z) \\ &= 3F_x^2(z) - 2F_x^3(z) = F_x^2(z)(3-2F_x(z)) \end{aligned}$$

and

$$f_y(z) = [6F_x(z) - 6F_x^2(z)]f_z(z) = 6f_x(z)F_x(z)(1 - F_x(z)).$$

(a) With $f_x(z) = \frac{1}{h}(|z| < \frac{h}{2})$, the uniform distribution with $\sigma_z^2 = \frac{h^2}{12}$, gives in $[-\frac{h}{2}, \frac{h}{2}]$:

$$F_y(z) = \frac{2}{h^3}(x + \frac{h}{2})^2(h - x), \; f_y(z) = \frac{6}{h^3}(x + \frac{h}{2})(\frac{h}{2} - xt) \quad \text{and} \quad \sigma_y^2 = \frac{h^2}{20}.$$

(b) With $f_x(z) = \frac{4}{h^2}(\frac{h}{2} - |z|)(|z| < \frac{h}{2})$ and $\sigma_x^2 = \frac{h^2}{24}$, we get

$$F_x(z) = \frac{2}{h^2}(x+\frac{h}{2})^2 - (z \in [-\frac{h}{2}, 0]) + (1 - \frac{2}{h^2}(x - \frac{h}{2})^2)(z \in [0, \frac{h}{2}]) + 1 \cdot (z > \frac{h}{2}),$$

and

$$\begin{aligned} F_y(z) \;=\; & \tfrac{4}{h^4}(z + \tfrac{h}{2})^4(3 - \tfrac{4}{h^2}(z + \tfrac{h}{2})^2), && (z \in [-\tfrac{h}{2}, 0]) \\ & (1 - \tfrac{2}{h^2}(x - \tfrac{h}{2})^2)^2(1 + \tfrac{4}{h^2}(x - \tfrac{h}{2})^2), && (z \in [0, \tfrac{h}{2}]) \\ & 1 && (z > \tfrac{h}{2}). \end{aligned}$$

Therefore, also

$$\begin{aligned} f_y(z) \;=\; & 3(z + \tfrac{h}{2})^3(1 - \tfrac{2}{h^2}(z + \tfrac{h}{2})^2)(\tfrac{2}{h})^4, && (z \in [-\tfrac{h}{2}, 0]) \\ & -3(z - \tfrac{h}{2})^3(1 - \tfrac{2}{h^2}(z - \tfrac{h}{2})^2)(\tfrac{2}{h})^4 && (z \in [0, \tfrac{h}{2}]) \end{aligned}$$

with $\sigma^2 y = \dfrac{23h^2}{20.78}$.

Comparing the transfer of distributions above with that of some simple filters can be done as follows. Let F be the $2n+1$ average filter and x a random sequence from a distribution $f_x(z)$, then the variable $s = \frac{x}{n}$ comes from the distribution

$$f_s(z) = nf_x(nz) = \frac{1}{h}f_x(\frac{z}{h}) \quad \text{, with } h = \frac{1}{n}.$$

If $\phi_k = f_x$ is a B-spline of order k, then the output of the average of two variables is

$$c(t) = \frac{1}{h^2}\int_{-\infty}^{\infty} \phi_k(\frac{x}{h})\phi_k(\frac{t-x}{h})dx = \frac{1}{h^2}\int_{-\infty}^{\infty} \phi_k(w)\phi_k(\frac{t}{h} - w)h\, dw,$$

so that $c(t) = \frac{1}{h}\phi_{2k}(\frac{t}{h})$, from the basic convolution properties of B-splines. Similarly the sum of m variables from $f_s(z)$ has the distribution

$$f_y(z) = \frac{1}{h}\phi_{nk}(\frac{z}{h}) = n\phi_{nk}(nz).$$

Comparing with the first example on M_1 we let $(Fx)_i = \frac{1}{3}[x_{i-1} + x_i + x_{i+1}]$. If $\phi_1^{(z)} = (|z| < \frac{1}{2})$, the output $y_i = \frac{1}{3}[x_{i-1} + x_i + x_{i+1}]$ is a sequence of numbers from $f_y(z) = 3\phi_3(3z)$, where ϕ_3 is the B-spline of order 3 given by;

$$\begin{aligned} \phi_3(x) \quad = \quad & \tfrac{1}{2}(x+\tfrac{3}{2})^2 \quad , \quad x \in [-\tfrac{3}{2}, -\tfrac{1}{2}] \\ & \tfrac{1}{2}(x-\tfrac{3}{2})^2 \quad , \quad x \in [\tfrac{1}{2}, \tfrac{3}{2}] \\ & \tfrac{3}{4} - x^2 \quad , \quad x \in [-\tfrac{1}{2}, \tfrac{1}{2}] \\ & 0 \qquad \text{otherwise.} \end{aligned}$$

From formulas for the moments of the splines, $\mu_2(3) = \frac{3}{12}$ and

$$\sigma_y^2 = \int_{-\infty}^{\infty} z^2 3\phi_3(3z)dz = \frac{1}{9}\int_{-\infty}^{\infty} x^2\phi_3(x)dx = \frac{1}{9}\mu_2(3) = \frac{1}{36}.$$

Thus the variance is reduced by a factor of 3, demonstrating the similarity in reduction of variance to that of M_1 which yields $\sigma_y^2 = \frac{1}{20}$. This is relatively easily generalized to $\dfrac{1}{2(4n+6)}$ for the smoother M_n, demonstrating the usual (slow) order of approximation of the standard deviation $\left(0\left(\frac{1}{\sqrt{n}}\right)\right)$ to zero.

When attempting to compare with the transfer of distributions of the $LULU$-operators we run into some computational complications, which can be witnessed in the simplest cases.

Let $x = \langle x_i \rangle$ be an i.i.d. with density f. Let $u = Ux_i = U_1x_i = \min\{m_j = \max\{x_{i-n+j}, \ldots, x_{i+j}\},\ j \in \{0, 1, \ldots, n\}\}$. Then

$$p(u \le z) = 1 - p(u > z) = 1 - p(m_j > z,\ \forall_j).$$

But $p(m_j > z, \forall_j) = p(x_i > z) + \varepsilon$. $\varepsilon = p(x_i \le z).p(x_\ell > z$, for some $i - n \le \ell < i)p(x_k > z, i < k \le i + n)$. Therefore

$$\begin{aligned} p(m_j > 2\forall_j) &= 1 - p(x_i \le z) \\ &\quad + p(x_i \le z)(1 - p(x_j \le z, \forall_{j<i}))(1 - p(x \le z, i < j \le i+n)) \\ &= 1 - F(x) + F(z)(1 - F(z)^n)^2. \end{aligned}$$

Therefore

$$F_u(z) = p(u \le z) = F(z)(2F(z)^n - F(z)^{2n}) = F(z)^{n+1}(2 - F(z)^n).$$

Differentiating this gives

$$\begin{aligned} f_n(z) &= f(z)(2(n+1)F(z)^n - (2n+1)F(z)^{2n}) \\ &= f(z)F(z)^n(2(n+1) - (2n+1)F(z)^n). \end{aligned}$$

Noting that $f_L(z) = f_U(-z)$ and integrating we get

$$\begin{aligned} F_L(z) &= \int_{-\frac{1}{2}}^{z} f_L(w)dw = \int_{-\frac{1}{2}}^{z} f_U(-w)dw \\ &= -\int_{\frac{1}{2}}^{-2} f_U(t)dt = -\int_{-\frac{1}{2}}^{-2} f_U(t)dt + \int_{-\frac{1}{2}}^{\frac{1}{2}} f_U(t)dt \\ &= 1 - F_U(-z). \end{aligned}$$

Thus the distributions of the output of L_1 can be simply obtained from those of U_1.

Example. Let f be the centered uniform distribution, then the output of U_1 has the distribution given by $f_u(z) = (z + \frac{1}{2})(2.5 - 3z)(x < \frac{1}{2})$. This is clearly biased in the positive direction with average value given by

$$\int_{-\infty}^{\infty} tf(t)dt = \frac{1}{12} \quad \text{and variance} \quad \int_{-\infty}^{\infty} t^2 f(t)dt = \frac{1}{15}.$$

The bias of U_n increases with n, with the average value given by $\dfrac{n^2}{2(n+1)(n+2)}$ which tends to $\frac{1}{2}$ as n grows. The variance tends to $\frac{1}{4}$.

Noting that the output of L_nU_n cannot be obtained from the composition of the individual mappings, since the output of U_n is increasingly correlated for increasing n, we have to go through probabilistic arguments that become increasingly tedious. Even for $n = 1$ it requires some work to yield,

$$\begin{aligned} F_{LU}(z) &= F(z)^2(1 + (1 - F(z))(1 + F^2(z)) \\ \text{and} \quad f_{LU}(z) &= f(z)F(z)(1 - F(z))(4 + F(z) + 5F(z)^2). \end{aligned}$$

These distributions still suggest an expected bias, which with the uniform f, can be calculated to yield an average of $\frac{1}{20}$ and a variance of $\sigma^2 = \frac{39}{400} - \frac{1}{21} \in \left[\frac{1}{21}, \frac{1}{20}\right]$.

This compares well with the variance of the median output. Intuition and simulation suggest that this equivalence will deteriorate as n grows. Seeking less bias, we often smooth with the smoothers C_n and F_n instead, which yield even more difficult calculations to yield a closed form expression for the distribution of the output. Though bounds may be obtainable, we might prefer simulation. Since real noise generally has to be experimentally characterized there seems to be no real motivation to explore much further. Initial simulations suggest that the variance reduction of the corresponding median compositions approximate those of C_n and F_n quite well, with a small bias remaining in the expected value. Further, and more systematic, evaluation of the transfer of distribution by $LULU$-operators has been done by Wild and others, but calculation becomes tedious.

Perhaps the more important statistical estimates for random error are those with the probability of pulses appearing in the resolution levels of a DPT. We consider again the response of the DPT to noise in the sequence $e = \langle e_i \rangle$, where

e_i is an i.i.d. random number from a probability density function f that is some B-spline (or in the limit, Gaussian).

An interesting question is what number of pulses can be expected in a particular resolution level, the expected amplitude and the variance. Taking the first stage of a DPT with $U_n L_n$ we consider only downward pulses in the first (highest-) resolution level. The pulses appear when $L_1 x_i > x_i$, that is when $\max\{x_{i-1}, x_{i+1}\} < x_i$.

$$p(\max\{x_{i-1}, x_{i+1}\} < z)k = F(z)^2, \quad \text{where} \quad F(z) = \int_{-a}^{z} f(t)dt.$$

The density function of $\max\{x_{i-1}, x_{i+1}\}$ is the derivative of $F(z)^2$, or $2F(z)f(z)$, and the distribution of $x_i - \max\{x_{i-1}, x_{i+1}\}$ is given by the convolute of f and $2Ff$.

Example. In the case where f is uniform and centered, $F(z) = (z - \frac{1}{2})(|z| \leq \frac{1}{2})$ and the convolute yields a distribution function given by $1 - z^2$ when $z > 0$. A downward pulse appears only when this distribution is positive. The average pulse therefore has an amplitude of

$$\frac{\int_0^1 t(1-t^2)dt}{\int_0^1 (1-t^2)dt} = \frac{1}{6}.$$

The total variation of the first part of the resolution layer, or $T(x - L_1 x)$, has an expected value of $2\frac{N}{3} \cdot \frac{1}{6}$, since the probability of such a pulse appearing is exactly $\frac{1}{3} = p(x_i > \max\{x_{i-1}, x_{i+2}\})$.

The negative pulses in the first resolution level can be handled similarly, but now it is clear that there is a lesser expectation, because of the following argument.

Let $z_j = (L_1 x)_j$. Then

$$\begin{aligned} x_i < U_1 L_1 x_i &= \min\{\max\{z_{i-1} z_i\}, \max\{z_i, z_{i+1}\} \Leftrightarrow z_{i-1}, z_{i+1} > x_i \\ &\Leftrightarrow \min\{x_{i-2}, x_{i-1}\}, \min\{x_{i+1}, x_{i+2}\} > x_i \Leftrightarrow x_{i-2}, x_{i-1}, x_{i+1}, x_{i+2} > x_i. \end{aligned}$$

The probability of this being true is precisely $\frac{1}{5}$, irrespective of the distribution f. Thus

$$P(U_1 L_1 x_i > x_i) = \int_{-\infty}^{\infty} F(z)^4 f(z)dz = \frac{1}{5}.$$

The total probability of a pulse appearing at a given point in the first resolution level is therefore $\frac{1}{3} + \frac{1}{5} = \frac{8}{15}$. Computing the average amplitude will result in a value less than $\frac{1}{6}$. Variance can also be computed. The total variation is thus also predictable. Simulations easily confirm the above calculations.

Clearly general expected values of amplitudes, variance, total variation and even probabilities of pulses appearing at a given resolution level are increasingly complicated to calculate. For practical problems the noise can however often be estimated by constructing histograms from simulation. Thus criteria for thresholding the first few levels to remove such noise with a reasonable probability can be fairly well selected and employed.

What is clear is that for such a sequence e more than half the pulses in the DPT appear in the first resolution level! (In a diadic wavelet decomposition this is almost half.) We can therefore expect a rapid lessening of the number of pulses and their amplitude in the higher resolution levels. The following graphs depict the expected values obtained by simulation of the fraction of the total variation remaining in the case of the uniform distribution. Various distributions yield similar results.

The modelling of noise in the case of selectors has some advantages and some disadvantages over the case of linear smoothers. Some key advantages seem to be the simple formula for the probability of a pulse appearing, the variation preservation and the associated full trend preservation. These may lead to some insight, simplification and estimation/bounding possibilities. This needs further research. For some background of noise modelling in image processing, Chapter 2 of the book of Starck et al. (3) can be consulted. What is clear however is that

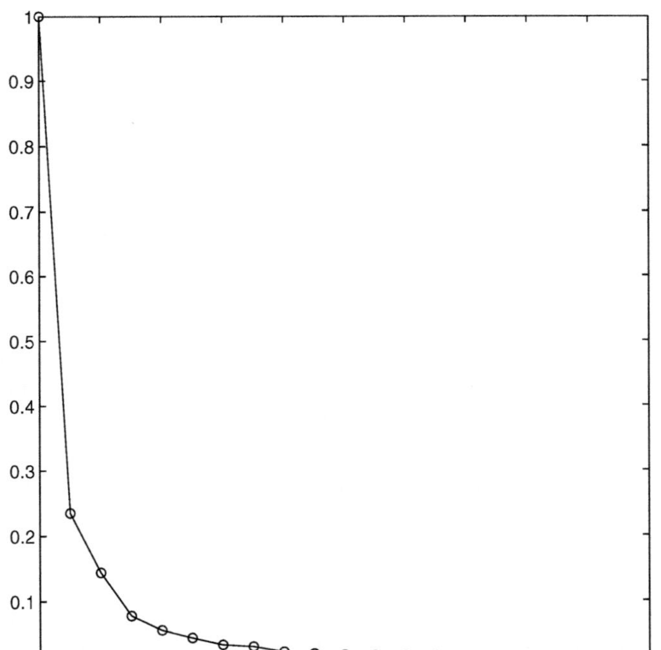

Figure 9.1. Graph of the fraction of the total variation left after peeling of the nth resolution level in the case of a uniform distribution.

even with linear Multiresolution Analysis, the approximation ultimately stored, or transmitted have quantizing and thresholding attached. Thus they also become nonlinear approximations of the original data.

Comparing linear and nonlinear smoothing in practice has another difficulty, since they are generally complementary procedures, and each has its merits. A simple example is the case of a contaminated sampling of a simple periodic function, where a Multiresolution Analysis with pulses reveals isolated large impulses as well as Gaussian noise. Using this diagnosis a simple low order *LULU*-based smoother (or median-) removes the large impulses and standard linear approximation yields a good reconstruction. An example is given in the figure below.

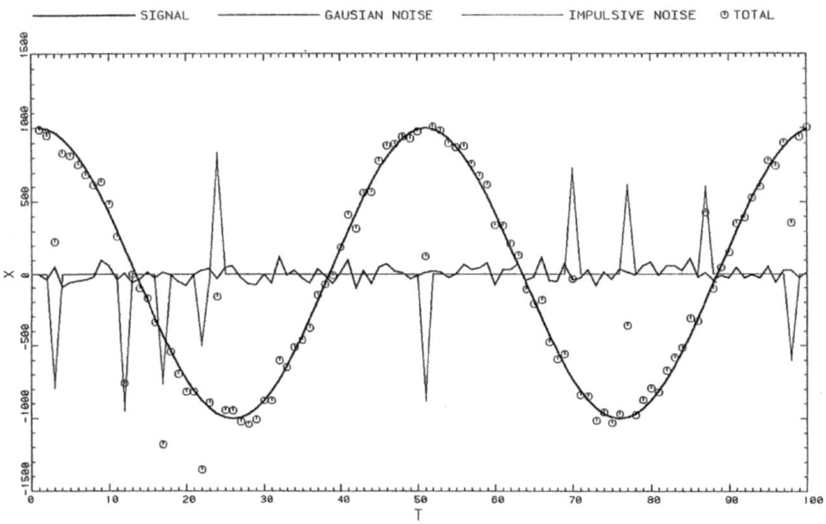

Figure 9.2. A simple reconstruction of a cosine function sampled with noise.

Going back to the first examples on page xi, we see that after analysis a morphological center of a *LULU*-operator with suitably large support leaves the trajectory with minimal damage available for further analysis. Linear smoothing attempts were not worth displaying.

The second example on page xii yields a fair comparison with small support linear filters and equivalent *LULU*-smoothers. The advantage was significant, and examples with L_nU_n with $n = 1, 2$ and 3 can be shown with those of equivalent averages.

The comparison here was entirely fair, although the author was on the boat when the measurements were taken and knew the intended speed. The boat was maintained at a constant speed of 35 knots (measured by other means) for a while, followed by a (linear) increase. This is clearly reflected in both the graphs, if interpreted by the intelligent human observer, whose descerning eye rejects the impulses still present. For a discussion of the essentially nonlinear nature of visual interpre-

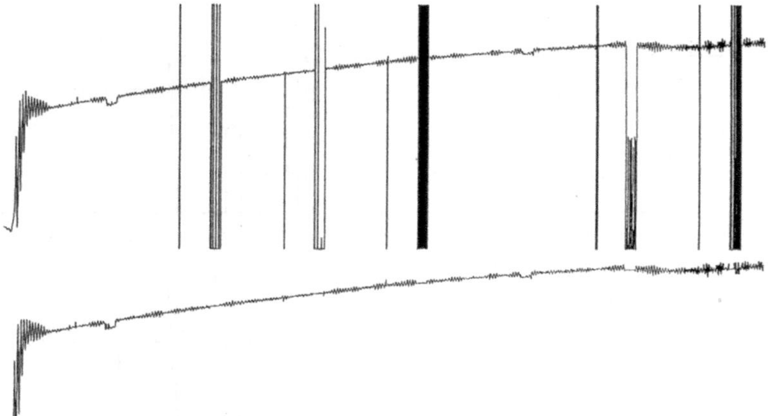

Figure 9.3. Decontamination of data with (real) impulsive noise.

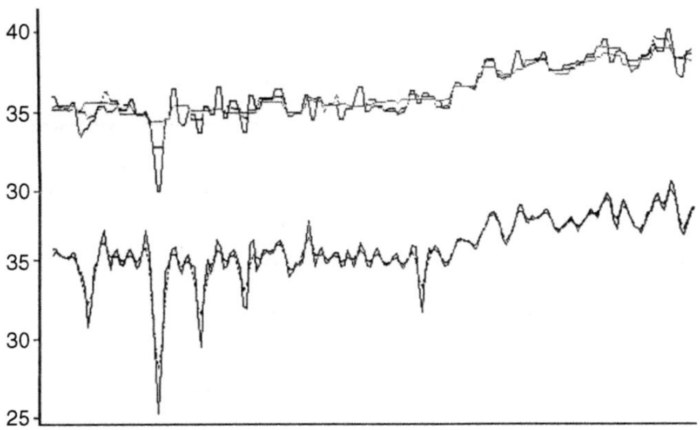

Figure 9.4. Comparison of nonlinear- and linear smoothers on speedometer data.

tation of the human brain, further reading is worthwhile. An example is the article of S. Mallet (12) in the further background reference. It is also worth considering how many simple optical illusions are constructable (18), whereas the ear seems less deceivable in such a simple fashion, how easily parallel lines are confused in the presence of another (monotone) line, and how exhausting and confusing (to say the least) a swaying horizon is at sea. Does the latter associate with the reference of the base as the outcome of successive *LULU*-smoothing, as in the final example?

As a final example of the properties of the pulse transform we consider again the *LULU*-decomposition of a sampled fort of Figure 7.2 distorted by noise (impulsive and Gaussian). It has the merit of demonstrating the economical coding of the essential recognizable shape (of C_5x) in four integers and four indexes, with code pulses appearing at some lower resolution levels, and the superior shape preservation compared to wavelet decompositions.

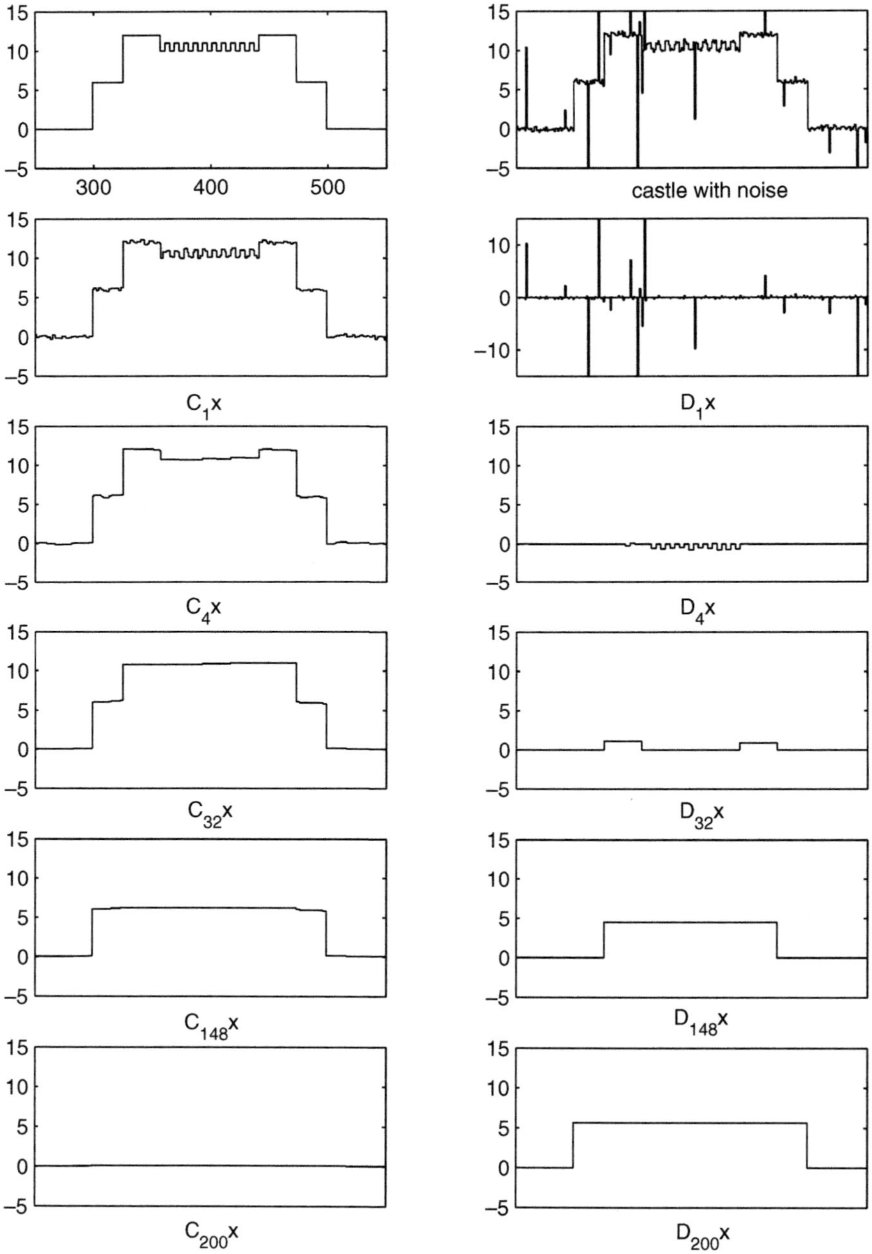

Figure 9.5. Graphs of impulsive noise and the sequences C_i and D_i for $i = 1$, 4, 32, 148, 200.

10. Interpretation and Future

Leg ich mich aus, so leg' ich mich hinein
Ich kann nicht selbst mein Interprete sein.
Doch wer nur steigt auf seiner eignen Bahn
Tragt auch mein Bild zu hellem Licht hinan.

Nietzsche

Hearing and Seeing are two physiologically different means of receiving information. Hearing and the Frequency Domain belong together, but in Vision the natural framework may be Mathematical Morphology.

Analyzing sequences for information with linear operators can be naturally associated with the Frequency Domain, since the eigensequences of such operators are periodic. Analyzing them for features like pulses and trends is different. Wavelet Theory has held promises but the constraints have become apparent. Mathematical Morphology, as a framework for analysis and interpretation, has been developed over several years, but has only recently been linked with other attempts at establishing mathematical structure. $LULU$-theory developed from the culture of intuitive association with the linear framework (of Electronic Engineers and Statisticians), but is a special part of Mathematical Morphology. Its perspectives yield a rich harvest of strong and intuitively simple results which the biased author risks calling beautiful.

This effort is intended to seduce mathematicians, scientists and engineers to explore this beautiful structure for interpretation and usefulness. It is deliberately written with passion rather than complete detail, an impressionistic painting rather than a photograph. The ideas have however grown out of practical problems tackled over more than two decades in several different environments. The simplicity, power and regularity of the practical successes has fed the passion that drove the analysis, for which the author never seemed or felt adequately prepared.

In attempting to interpret what $LULU$-theory has yielded, the following can be claimed.

1. A framework for analysis and comparison of several useful nonlinear smoothers has been established that contributes to the selection, design and understanding of smoothers for a specific purpose.

2. Conceptual links have been established with the existing (linear) framework for analysis, comparison and design of smoothers, as well as Mathematical Morphology.
3. A concept of "Smoothness" and "Trend" (Local Monotonicity) has been established and linked to an established natural measure of smoothness (Total Variation).
4. A Concept of "Pulse" has been established that is complementary to that of "Trend".
5. Fundamental operators (C_n and F_n) have been defined and investigated for establishing a class of (nonlinear-) Multiresolution Analysis that decompose a sequence for such "Trend" and "Pulse" content.
6. The attributes of Neighbor Trend Preservation (ntp) and Full Trend Preservation (ftp) and that of Co-idempotence have been defined and demonstrated to be as useful and important as the well-known concepts of Idempotence and Syntoneness.
7. Heuristically derived procedures (Median Transforms) in Image Processing, which have been known to work well (but lacking theory), have had their success supported by theory and the perceived superiority over Morphological Filters has been shown to be due to lack of fair selection.
8. Heuristic truths have emerged that have often resulted in good results in particular applications in Science and Industry. Involvement with several partners in Laser Spectroscopy, Positioning Problems, Earthquake Research, Image Processing, Financial Analysis, as well as Research in Mathematics and Statistics, is ongoing and growing beyond expectation.
9. Relevance (value) of some mathematical structures has been exposed. Mathematical research into these should yield many simpler arguments and clearer insights.
10. A Discrete Pulse Transform (DPT) has been established in analogy to a Discrete Fourier Transform (DFT). The primary use may be in image processing. A remarkable (and unexpected) consistent behavior has been proved and indications for further consistency presented.
    Unlike the DFT the resolution levels are not subspaces and there is no hope, nor justification, for expecting consistent decomposition of arbitrary linear combinations of sequences from such resolution levels. If however a specific sequence is decomposed, then those specific resolution levels can be used to choose an arbitrary linear combination with non-negative coefficients, and have guaranteed consistent decomposition. Noting that there is no negative luminosity in image processing, this seems sufficient for the purpose at hand, and as much as can be expected. Elementary modifications of such a decomposition can be stored and transmitted, and the reconstructed image will be predictably decomposed by the DPT, including the strong consistency formulated as "The Highlighting Conjecture".

11. Since the manuscript has been delivered to the publisher, some additional research has yielded results on some diverse lines. The interested reader can consult the list for further background reading for some published results. An article, written long ago with Malkowsky, introduces an analogy in Real Analysis. The article on quasi-inverses links up with Chapter 5, and shows that the $LULU$-operators and compositions are optimal in damage minimization amongst general min-max operators. Included are some basic results on further links in $LULU$-theory, like a simple proof that the median operator M_1 is Variation Reducing, but arbitrarily slowly. Further advances in this direction have been submitted. Two further articles on applications to Financial Data, and on the transfer of distributions have appeared. Two further articles are being prepared on different aspects of the consistency of the DPT. Prospects are good that one will contain a proof of the Highlight Conjecture.

Future. The author has exposed his ideas in the hope that contact, debate and criticism can lead to more understanding of his favorite research area. It is hoped that more competent researchers can solidify and extend the ideas that have been behind this effort. If only a few of the avenues of research that have been opened yield some result, the author would be grateful.

References

[1] E. Ataman, V.K. Aarte and K.M. Wong, *A Fast Method for Real-Time Median Filtering,* IEEE Trans. Acoust. Speech Signal Process. **ASSP-28**, No. 4 (1980).

[2] E. Ataman, V.K. Aarte and K.M. Wong, *Some Statistical Properties of Median Filters,* IEEE Trans. Acoust. Speech Signal Process. **ASSP-29**, No. 5 (1981).

[3] A. Bijaoui, F. Murtagh, J.L. Starck, *Image Processing and Data Analysis – The Multiscale Approach.* Cambridge University Press, 1998.

[4] A.C. Bovic, A. Restrepo, *Locally Monotonic Regression*, IEEE Transactions on Signal Processing. Sept. 1993, Vol. 41, No. 9.

[5] A.R. Butz, *A Class of Rank Order Smoothers,* IEEE Trans. Acoust. Speech Signal Process. **ASSP-34**, No. 1 (1986).

[6] A.R. Butz, *Some Properties of a Class of Rank Order Smoothers*, IEEE Trans. Acoust. Speech Signal Process. **ASSP-34**, No. 3 (1986).

[7] N.C. Callagher and G.L. Wise, *A theoretical analysis of the properties of median filters,* IEEE Trans. Acoust. Speech Signal Process. **ASSP-29**, No. 6 (1981).

[8] C.K. Chui, *Wavelets: A Mathematical Tool for Signal Analysis.* SIAM, Philadelphia, 1997.

[9] E. Cloete, B. Opperman, *An Asynchronous parallel computing algorithm for a non-linear filter:* Proc. 18th SANUM Conference, Durban, 1992.

[10] L. Collatz, *Monotonicity in Numerical Mathematics,* Proc. 1st SANUM Conference, Durban 1973.

[11] R.W. Hamming, *Digital Filters* Prentice-Hall, Englewood Cliffs, N.Y., 1977.

[12] R.L. Hamming, *The Frequency Approach to Numerical Mathematics in Studies in Numerical Analysis,* Editor: B.K.P. SCAIFE, Academic Press.

[13] C. Lanczos, *Applied Analysis,* Prentice Hall, N.J., 1956.

[14] C.L. Mallows, *Some Theory of Nonlinear Smoothers,* Ann. Statist. **8**, No. 4 (1980), 695–715.

[15] P. Maragos, R.W. Schafer, *Morphological Filters - Part II: Their relations to Median, Order-Statistic, and Stack-Filters,* IEEE Transactions on Acoustics, Speech and Signal Processing, Vol. **ASSP-35** (1987), 1170–1184.

[16] A.E. Marquardt, L.M. Toerien and F. Terblanche, *Applying nonlinear, Smoothers to Remove Impulsive Noise from Experimentally sampled data,* R and D. Journal, SA I Mech E. Vol. 7, No. 1 (1991).

[17] T.A. Nodes and N.C. Callagher, *Median Filters: Some Modifications and Their Properties*, IEEE Trans. Acoust. Speech Signal Process. **ASSP-35**, No. 5 (1982).

[18] K.R. Popper and J.C. Eccles, *The Self and its Brain,* Springer International, 1981.

[19] C.H. Rohwer, *Idempotent One-Sided Approximation of Median Smoothers*, Journal of Approximation Theory. Vol. 58, No. 2 (1989), 151–163.

[20] C.H. Rohwer, L.M. Toerien, *Locally Monotone Robust Approximation of Sequences*, Journal of Computational and Applied Mathematics **36** (1991), 399–408.

[21] C.H. Rohwer, *Projections and Separators*, Quaestiones Mathematicae **22** (1999), 219–230.

[22] C.H. Rohwer, *Fast Approximation with Locally Monotone Sequences.* Proceedings 4th FAAT Conference, Maratea. Supplemento ai rendiconti del Circolo matimatico di Palermo, Series II numero 68 – anno 2002.

[23] C.H. Rohwer, *Variation Reduction and LULU-smoothing*, Quaestiones Mathematicae **25** (2002), 163–176.

[24] C.H. Rohwer, M. Wild, *Natural Alternative for One Dimensional Median Filtering.* Quaestiones Mathematicae. **25** (2002), 135–162.

[25] C.H. Rohwer, *Multiresolution Analysis with Pulses*, Advanced Problems in Constructive Approximation (Editors: M.D. Buhmann, D.A. Mache), International Series of Numerical Mathematics Vol 142, 165–186, Birkhäuser Verlag Basel, 2002.

[26] C.H. Rohwer, *Fully Trend Preserving Operators.* Quaestiones Methematicae **27** (2004), 217–229.

[27] H.L. Royden, *Real Analysis*, Macmillan, N.J., 1969.

[28] I.J. Schoenberg, *On Spline Functions, in Inequalities.* (Editor: O. Shisha) Academic Press New York, 1967.

[29] J. Serra, *Image Analysis and Mathematical Morphology*, Acad. Press London 1982.

[30] J.W. Tukey, *Exploratory Data Analysis*, Addison-Wesley, Reading, MA, 1977.

[31] P.F. Velleman, *Robust Nonlinear Data Smoothers: Definitions and Recommendations,* Proc. Natl. Acad. Sci. USA, **74**, No. 2 (1977), 434–436.

[32] M. Wild, *On the Idempotency and Co-idempotency of the Morphological Centre*, International Journal of Pattern Recognition and Artificial Intelligence, Vol. 15, No. 7 (2001), 119–1128.

[33] Zhou Xing-Wei, Yang De-Yun, Ding Run-Tao, *Infinite Length Roots of Median Filters,* Science in China, Ser. A, Vol. 35 (1992), 1496–1508.

Some further background

[1] G.R. Arce and N.C. Gallagher, *Jr., State description for the root signal set of median filters*, IEEE Trans. Acoust. Speech Signal Process. **ASSP-30** (1982), 894–902.

[2] G.R. Arce and M.P. McLoughlin, *Theoretical analysis of the max/median filter*, IEEE trans. Acoust. Speech Signal Process. **ASSP-35** (1987), 60–69.

[3] N. Balakrishman and C.R. Rao (eds), *Handbook of Statistics*, Elsevier Science B.V. 1998.

[4] C.K. Chui, *An introduction to Wavelets*, Academic Press, Onlander. 1992.

[5] R. Coifman, F. Geschund and Y. Meyer, *Applied and computational analysis* **10**, 27–44, (2001).

[6] W.J. Conradie, T. de Wet, M.D. Jankowitz, *An Overview of LULU Smoothers with Application to Financial Data.* S.E.E. (To appear).

[7] W.J. Conradie, T. de Wet, M.D. Jankowitz, *Exact and Asymptotic Distribution of LULU Smoothers.* Journal of Computational and Applied Mathematics. (To appear).

[8] J.P. Fitch, *Software and VLSI algorithms for generalized ranked order filtering,* IEEE Trans. Circuits and Systems. **CAS-34** (1987), 553–559.

[9] K.J. Hahn and K.M. Wong, *Median filtering of spectra*, Proc. IEEE International Electrical and Electronic Conference, Toronto, Sept. 1983, pp. 352–355.

[10] O. Kao, *Modification of the LULU operators for preservation of critical image details*, The 2001 International Conference on Imaging Science, Systems and Technology, Las Vegas, Nevada, USA.

[11] E. Malkowsky, C.H. Rohwer, *The LULU-semigroup for Envelopes of Functions*, Quaestiones Mathematicae **26** (2004).

[12] S. Mallet, *Applied mathematics meets signal processing*, DOC. MATH. Extra volume ICM. I. 319–338 (1998).

[13] M.P. McLoughlin and G.R. Arce, *Deterministic properties of the recursive separable median filter*, IEEE Trans. Acoust. Speech Signal Process. **ASSP-35**, (1987), 98–106.

[14] P.M. Narendra, *A separable median filter for image noise smoothing,* IEEE Trans. Pattern Anal. Machine Intelligence. **PAMI-3** (1981), 20–29.

[15] T.A. Nodes and N.C. Gallagher, *Jr., Two-dimensional root structures and convergence properties of the separable median filter,* IEEE Trans Acoust. Speech Signal Process. **ASSP-31** (1983), 1350–1365.

[16] L.R. Rabiner, M.R. Sambur and C.E. Schmidt, *Applications of a nonlinear smoothing algorithm to speech processing,* IEEE Trans, Acoust. Speech Signal Process. **ASSP-23**, No. 6 (1975).

[17] C.R. Rao, *Has statistics a future? If so, in what form?* Submitted World Scientific, 2002.

[18] C.H. Rohwer, *Quais-inverses and Approximation with min-max Operators,* Proceedings 5th FAAT Conference, Maratea. Supplemento ai rendiconti del Circolo matematico di Palermo. (To appear).

[19] U.E. Ruttimann and R.L. Webber, *Fast computing median filters on general purpose processing systems,* Optical Engineering, **25**, No. 9 (1986).

[20] A.F. Siegel, *Robust regression using repeated medians,* Biometrika, **69**, 1 (1982), 224–242.

[21] R.N. Strickland and M.R. Gerber, *Estimation of ship profiles from a time sequence of forward-looking infrared images*, Optical Engineering. **25**, No. 8 (1986), 995–1000.

[22] M. Wild, *Idempotent and co-idempotent stack filters and min-max operators*, Theoretical Computer Science **299** (2003), 603–631.

Index

GPSR Compliance

The European Union's (EU) General Product Safety Regulation (GPSR) is a set of rules that requires consumer products to be safe and our obligations to ensure this.

If you have any concerns about our products, you can contact us on ProductSafety@springernature.com

In case Publisher is established outside the EU, the EU authorized representative is:

Springer Nature Customer Service Center GmbH
Europaplatz 3
69115 Heidelberg, Germany

Batch number: 10578049

Printed by Printforce, the Netherlands